*The Darien Journal of
John Girardeau Legare,
Ricegrower*

The Darien Journal
of
John Girardeau Legare, Ricegrower

EDITED BY
BUDDY SULLIVAN

The University of Georgia Press
Athens & London

Published in 2010 by the University of Georgia Press
Athens, Georgia 30602
www.ugapress.org
© 1997 by Buddy Sullivan
Preface to the 2010 Edition © 2010 by the University of Georgia Press
All rights reserved
Designed by Buddy Sullivan
Set in Goudy Oldstyle

Printed digitally in the United States of America

Library of Congress Cataloging-in-Publication Data

Legare, John Girardeau, 1852–1932.
The Darien journal of John Girardeau Legare, ricegrower /
edited by Buddy Sullivan.
p.   cm.
Originally published: Darien, Ga. : B. Sullivan, 1997.
Includes bibliographical references and index.
ISBN-13: 978-0-8203-3560-5 (hardcover : alk. paper)
ISBN-10: 0-8203-3560-6 (hardcover : alk. paper)
1. Legare, John Girardeau, 1852–1932 — Diaries.
2. Rice farmers — Georgia — Darien — Diaries.
3. Plantation life — Georgia — Darien — History.
4. Community life — Georgia — Darien — History.
5. Darien (Ga.) — Biography.
6. Darien (Ga.) — Social life and customs.
7. Darien (Ga.) — History. I. Sullivan, Buddy. II. Title.
F294.D26L44 2010
975.8'737 — dc22   2010007673

British Library Cataloging-in-Publication Data available

The Darien Journal of John Girardeau Legare, Ricegrower
was originally published in 1997 by Buddy Sullivan
through a grant from the city of Darien.

Frontispiece: Photograph of John Girardeau Legare.
Courtesy of Dr. Mary Sue Murray.

This one is for Amanda

Contents

Preface to the 2010 Edition
ix

Editor's Introduction:
The Darien World of John Girardeau Legare
1

John Girardeau Legare's Journal
25

Notes
113

Index
151

Preface to the 2010 Edition

The small Georgia coastal town of Darien epitomized the industrial post–Civil War "New South" long before Henry W. Grady, progressive editor of the *Atlanta Constitution*, popularized the term in the 1880s. Darien actually knew no Reconstruction travails, at least economically, relative to the rest of the devastated South after the war. Instead, it experienced a remarkable commercial resurgence that resonated to the economic hum of sawmill gang saws and steamboat engines, all energized by a propitious infusion of northern capital. There thus developed a postbellum prosperity in Darien that benefited both blacks and whites alike.

Paradoxically, native South Carolinian John Girardeau Legare (pronounced Lagree), protagonist of the present study, personified the "Old South," particularly when set amid the energetic new industrial economy of Darien. Legare was an agriculturist by profession, a member of a gradually dwindling rice-planting class in an era when such avocations were more and more a vanishing aspect of the tidewater commercial South.

These considerations notwithstanding, Legare's Darien journal, the highly interesting document contained herein, can tell us much about postbellum life in Darien, typical of a small, vibrant southern industrial town. Not only does it instruct us on the postbellum rice and timber economy of Darien and the Altamaha River basin, it also illuminates our understanding and appreciation of the often peculiar circumstances that shaped people's lives in the half century from 1875 to 1925, especially in regard to the social mores, race relations, and medical, scientific, and technological developments of the time.

The first edition of the Legare journal was published in 1997 through a grant from the city of Darien. This new edition by the University of Georgia Press makes the journal available to a wider audience of researchers, scholars, and others interested in both southern and coastal Georgia history. Since its original publication, the scholarship on the rice culture of the Georgia and South Carolina tidewater has greatly expanded, as has the availability of literature on specific aspects of postbellum and early-twentieth-century coastal life.

In my own research, I owe much to the outstanding exposition of Mart A. Stewart in *What Nature Suffers to Groe: Life, Labor and Landscape on the Georgia Coast, 1680–1920* (Athens: University of Georgia Press, 1996). This unique study compellingly elucidates the critical link between the local ecosystem—and its attendant environmental factors—and the human activities and economy of the coast. Stewart's pathbreaking work certainly provided the impetus for my own subsequent investigations and findings within this same arena, most recently posited in my paper "Ecology as History in the Sapelo Island National Estuarine Research Reserve" (Sapelo Island, Ga.: Sapelo Island NERR, 2008).

Two outstanding recent studies, while focusing on antebellum rice and slave culture, are nonetheless highly useful to an understanding of the present subject. These are *Them Dark Days: Slavery in the American Rice Swamps* by William Dusinberre (repr., Athens: University of Georgia Press, 2000) and *Dwelling Place: A Plantation Epic* by Erskine Clarke (New Haven, Conn.: Yale University Press, 2005). The late Malcolm Bell, my mentor and friend, two decades ago produced the definitive, and lasting, study of nineteenth-century rice culture and the lives of slaves, slave masters, and freedmen in tidewater Georgia, *Major Butler's Legacy: Five Generations of a Slaveholding Family* (Athens: University of Georgia Press, 1987). No investigation of the subject would be complete without consulting that thorough and illuminating volume. The postbellum rice industry is described in two contemporaneous accounts from the late nineteenth century, *Seed from Madagascar* by Duncan Clinch Heyward (repr., Columbia: University of South Carolina Press, 1993) and *A Woman Rice Planter* by Elizabeth Allston Pringle (repr., Columbia: University of South Carolina Press, 1992). For the rice culture of the Altamaha delta, amid which Legare conducted his own agricultural activities, useful accounts are James E. Bagwell, *Rice Gold: James Hamilton Couper and Plantation Life on the Georgia Coast* (Macon, Ga.: Mercer University Press, 2000); Buddy Sullivan, *"All Under Bank": Roswell King, Jr., and Plantation Management in Tidewater Georgia, 1819-1854* (Darien, Ga.: Liberty County Historical Society, 2003); Albert V. House, ed., *Planter Management and Capitalism in Ante-bellum Georgia; The Journal of Hugh Fraser Grant, Ricegrower* (New York: Columbia University Press, 1954); Theresa A. Singleton, ed., *The Archaeology of Slavery and Plantation Life* (Orlando, Fla.: Academic Press, 1985); and two essential contemporary accounts, Frances Anne Kemble, *Journal of a Residence on a Georgian Plantation in 1838-1839* (repr., Athens: University of Georgia Press, 1984) and Frances Butler Leigh, *Ten Years on a Georgia Plantation since the War* (repr., Savannah, Ga.: Library of Georgia, Beehive Foundation, 1992). The latter two books are set on Butler's Island near Darien.

On postbellum race relations, including those in the Darien world of John G. Legare, W. Fitzhugh Brundage, *Lynching in the New South: Georgia and Virginia, 1880-1930* (Urbana: University of Illinois Press, 1993), is another essential reference. This important study provides useful insights regarding race in Georgia in the postbellum period, with particular reference to the Darien "insurrection" of 1899. Legare was greatly concerned with the meteorological events of his time and in this regard, Walter J. Fraser's *Lowcountry Hurricanes: Three Centuries of Storms at Sea and Ashore* (Athens: University of Georgia Press, 2006) is invaluable. The backdrop of Darien and McIntosh County history attendant to the present study is detailed in my *Early Days on the Georgia Tidewater: The Story of McIntosh County and Sapelo*, 6th ed. (Darien, Ga.: McIntosh County Board of Commissioners, 2001). My recent study *High Water on the Bar* (Darien, Ga.: Downtown Development Authority, 2009) places Darien's late-nineteenth-century history in context, centering on the town's lumber industry and maritime shipping trade.

In transcribing Legare's journal, I have elected to fully, and carefully, replicate the text of the original, following Legare's spelling, syntax, and punctuation exactly as he wrote it. Information within brackets is occasionally utilized where amplification on a journal entry is obviously needed. The endnotes provide biographical details and source materials consulted for most of the persons and places cited by Legare. In the introduction I have attempted to place Legare and his Darien world in context, from the mid-1870s to the early 1930s. The information therein should be helpful to those utilizing the document who are not familiar with the nuances of local life and history during the period covered. This essay also serves as a brief biography of Legare himself, defines his South Carolina roots, and establishes his impact, which was substantial, upon the civic, economic, and religious aspects of life in the Darien community.

In acknowledgement, I wish to thank once again for her support and encouragement Dr. Mary Sue Murray of Douglasville, Georgia, a descendant of John Girardeau Legare and custodian of her antecedent's original journals. Also, I thank the late Annie Fisher Gill of Darien, whose vast archive of genealogical materials provided a rich source of information in compiling the data in the notes. I wish additionally to thank Nicole Mitchell, director of the University of Georgia Press, for her editorial guidance and support, and her administrative assistant, Margaret Swanson, for her efficiency and encouragement—and patience—in facilitating many of the logistical details for this new edition of Legare's journal. Finally, thanks go to the Press's anonymous reader who reviewed the first edition and provided many useful editorial comments and suggestions.

*The Darien Journal of
John Girardeau Legare,
Ricegrower*

EDITOR'S INTRODUCTION

The Darien World of John Girardeau Legare

The old rice aristocracy of tidewater Georgia and South Carolina died in the ashes of civil war. The great rice plantations which prospered on the engines of huge labor forces of slaves, enormous outlays of capital and the ingenuity and resourcefulness of a handful of planters had represented the greatest concentration of wealth the antebellum South had ever seen when the national conflict of 1861-65 effectively and forever sealed the end of both an era and a way of life. It was no exaggeration when it was said that the crescent of south Atlantic seaboard from Georgetown District, South Carolina, southward to Glynn County, Georgia, was known as the Rice Kingdom of the world. No region produced more rice prior to the Civil War. And because of the enormous value of this section to the Confederacy, no region of the South felt the direct impact of war more than this one. When the blood-letting was done, the old rice planting aristocracy was obliterated—the slaves emancipated, the plantations destroyed, the wealth gone.

The planter class, what remained of it, had little left but their land. For some that was enough to try to begin anew but, as Peter Coclanis appropriately comments, it was but "the shadow of a dream." A few of the younger members of the former aristocracy tried to resurrect the rice industry, becoming burdened with debt in the process and struggling for three decades to recapture at least a measure of the wealth and grandeur that their fathers had labored so mightily to sustain. The hardier ones held on until just after 1900 until the realities of labor shortages, lack of adequate capital, declining markets and the vagaries of the coastal storm season finally sealed the Atlantic rice industry's fate. John G. Legare of Darien was one of those who made the attempt to keep rice cultivation alive in the "kingdom" but, as his personal Journal increasingly reveals, through his thoughts and observations, he was among the first to realize that rice would never again be the valuable commodity on the south Atlantic tidewater that it had once been. He correctly reasoned sooner than most that the industry, irrevocably, had to end. It could not sustain itself indefinitely. When he realized that there were no more profits to be had in rice, however small, he walked away from it and never looked back. Others continued for a few more years, always believing that, somehow, the next crop would be the one which started them back on the road to recovery.

As one traverses the coastal highway through the brackish tidal marshes, a few lingering vestiges of this legacy remain. Abandoned rice fields and irrigation ditches are now havens for migratory waterfowl and the intricate systems of canals and rice dikes remain as visible symbols of a way of life that has forever disappeared. The tidegates are rotting away in the levees, no longer able to keep the flooding and ebbing tides out of the square fields where rice once thrived. The old threshing and winnowing houses, mills, barns and workers quarters have long since disappeared, leaving only a few crumbling brick remains and the occasional rice mill chimneys which still stand watch as silent sentinels over the remnants of the former rice culture. But even more than that, it is the moss-draped live oaks, the drooping willows on the dikes and the huge expanses of marshes which still change color with the seasons, which remain as the most enduring and tangible testimony to the days of tidewater rice cultivation. These will always be here, as will the denizens of their habitat.

John Girardeau Legare of South Carolina was only twenty-five years old when he made the town of Darien, McIntosh County, Georgia, his adopted home in 1877. He probably little realized it then but this small coastal community would be his permanent residence for the remaining fifty-five years of his life. As so amply demonstrated by his Journal, Legare was an interesting man who lived in interesting times. He came to tidewater Georgia practically without a penny in his pocket, and with a young family to feed and care for. The southeastern coast was still recovering from the ravages of war—few places had suffered more than Darien. On the surface of it, a place such as Darien, Georgia seemed to offer poor prospects for a young family seeking a new beginning.

Through sheer dint of hard work and energy, along with a special blend of talent and resourcefulness that typified his life, particularly during his Darien years, Legare made a living for himself and his family that, while far from lavish or ostentatious, was nonetheless comfortable, stable and respectable. For more than twenty-five years Legare was a rice grower, a fast-vanishing breed during the postbellum years when rice-planting was not nearly as lucrative a profession as it had been in pre-war times.[1] Legare managed and administered rice operations under contract to a series of land owners in the rich, alluvial Altamaha delta of McIntosh County, with his primary activities centered on Generals and Champneys islands and, for a time, the well-known Butler's Island. Rice cultivation in these times was difficult. Only a handful of planters in McIntosh County and neighboring Glynn made the attempt. Rice, once the king of tidewater cash crops was, by Legare's time, in the twilight years of its dominance as an agricultural staple in coastal Georgia and South Carolina.[2]

Shortly after the turn of the twentieth century Legare, like most of his contemporaries who were similarly engaged, gave up the increasingly unprofitable efforts in rice and found new professional challenges as a local administrator. For almost three decades Legare served both McIntosh County and the City of Darien as what was tantamount to manager of day-to-day operations. He had the perspicacity to remain segregated from direct involvement in local politics—he ran for public office in Darien only once, and in that endeavor he was uniformly unsuccessful, a result which he accepts and philosophically records in his journal with typical candor.

Legare's life spanned a period of Darien's most interesting history. Although he plainly did not realize it at the time, Legare was unique in that he both experienced and witnessed some of the most exciting, and unusual, moments in the annals of Darien—certainly some of the town's most volatile events. During Legare's half century of residence in Darien and of his diligent recording of the minutiae of daily life there, his beloved town experienced more than a dozen major conflagrations; survived the ravages of two direct strikes by hurricanes only two years apart (1896 and 1898), and even an earthquake (1886); a serious race riot in 1899, and a meteoric economic cycle reflected in one of the most unusual boom-to-bust commercial swings any town its size has ever experienced.

Darien's commercial lifeblood flowed around timber and rice, and mainly the former. For a time, Darien was the most productive timber port on the south Atlantic coast; and rice shipments from Darien, though far from matching those of antebellum times, were nonetheless comparable with the most productive on the coast. But both industries, timber and rice, died hard and expired quickly in and around Darien. In a ten-year period from 1901 to 1911 the town rapidly went from peak to valley in terms of commercial shipments of lumber, naval stores and agricultural products. It had long been understood that since the end of slavery the cultivation of rice was no longer a profitable venture on the tidewater. The lumber industry died when the pine timber ran out upriver—along the Altamaha, the primary conveyor of millions of board feet of pine to the active sawmills of Darien. Timber died with little warning and with startling rapidity, leaving a stunned Darien citizenry in its wake which was slow to realize that nothing would ever be the same again.

Legare saw many of these developments coming—indeed the salient point in his journals during the years of decline were his views on the need for the town to diversify its economic potential, and pleas for the city fathers to look beyond the decline and rescue the commercial lifeblood of Darien, pleas which, as events would prove, went largely unheeded. Legare's vision of decline fell on the deaf ears of a Darien populace spoiled by the glut of yellow pine timber rafted down the Altamaha to the town's mills and the international shipping waiting to take the timber to points all over the globe.

Legare began keeping his journal, a written record of the events of his life, his family's and that of Darien, not long after he came to the Georgia tidewater in 1877. But even before that, the young South Carolinian had already accumulated a lifetime of experiences compared to most. Born on the 26th of April, 1852 in Adams Run, South Carolina, Legare was the scion of a family with a long and sweeping heritage sustained by generations of influence and impact upon the society, business and commerce of Charleston and the Carolina lowcountry. Legare's introduction to his journal is instructive, for it describes a normal boyhood as a lad of the waning antebellum years in the small, dusty, mainland town of Adams Run, a short distance southwest of Charleston and west of nearby Edisto Island. However, his adolescence and coming of age were anything but normal—they were shaped by war and Reconstruction, and were unquestionably traumatic for any growing youth of the times. The war and its bitter aftermath altered his perceptions and vision completely and forever, sentiments clearly borne out in his "Introductory," which was written some years after the events occurred with the benefit of wisdom and hindsight. Just short of being nine years old when the first shots were fired on Fort Sumter in nearby Charleston harbor, Legare was too young for Confederate military service, but the horrors of war were brought home to him, nonetheless. As the fighting neared its conclusion, the lad of twelve experienced first-hand the terror of people dying violently and a city being blasted to pieces around him. "Saturday, Feby. 18th, 1865—the most awful day of my life, the experiences of that day will not be forgotten," Legare observes in his compelling account of those hours he and his family huddled in a dark building as Union shells exploded nearby and retreating Confederate forces burned cotton warehouses and ordnance facilities as they evacuated Charleston.[3]

During the difficult years of Reconstruction, the young Legare applied himself well and learned a number of crafts which would prove exceedingly useful to him later in life—he mastered the rudiments of rice cultivation during this time, while also serving as a journeyman carpenter. The benefits Legare gained from his improving skills as a carpenter and a brief stint as a sawmill operator are often reflected through the pages of his Journal after his move to Darien. On November 24, 1874, at the age of twenty-two, Legare married Charlotte Smith Hamilton, a widow four years his senior. Legare gained a stepson by the union, five-year old Arthur St. Clair Hamilton. Finding the prospects for steady employment to be rather tenuous in the environs of Charleston, Legare, in the fall of 1877, accepted an offer from a family friend and fellow South Carolinian, Nathanial H. Barnwell of Glynn County, Georgia, to move to that section and assist in the operation of various local rice planting operations of which he was manager. It was through this connection that Legare moved his young family into what soon developed into a promising and permanent situation in the Altamaha rice delta. At the start of 1878, Legare took charge of the Generals Island rice plantation for Barnwell. Since Generals Island was situated just across the north branch of the Altamaha from the small town of Darien, it followed that Legare and his family make that community their new home.

The Darien of the 1870s was a town that had risen from the ashes of war—in 1863 the town, then an insignificant cotton port with barely 500 residents, had been put to the torch by invading Union forces stationed on nearby St. Simons Island in one of the more controversial incidents of the Civil War.[4] But within a decade of the war, Darien was a booming timber town. Local sawmills were processing millions of board feet of yellow pitch pine timber rafted down the Altamaha River from the southeast Georgia pine barrens and the town was rapidly gaining an international flavor due to the diversity of shipping arriving for lumber from Europe and South America.

It was the river which was easily recognized as the lifeblood of Darien. The river was the conveyor for timber coming to the town's mills and it was the river which ebbed and flowed to alternately flood and drain the rice fields as the old industry struggled to reassert itself following the sudden loss of capital and free labor as a result of the war. Few understood the importance of the river to Darien's survival as did Richard Grubb, the young editor of the local weekly newspaper, the *Darien Timber Gazette*, so named because of the town's heavy reliance on timber as the prime commodity of its economic salvation. "Dick" Grubb had revived the town's newspaper in the spring of

1874 and it was not long before the paper "became an indispensable medium of news and comment about commerce and culture on Georgia's coast."⁵ Grubb's views regarding the significance of the Altamaha are interesting in that they reflect an almost global perception in comparison with another, more noteworthy, locale:

> "The Nile is said to be everything to Egypt [Grubb wrote]; in fact without the Nile, there would be no Egypt, only a continuation of the sands of the Desert. To a certain extent the same may be truly said of the Altamaha River and Darien. Without the Altamaha, there would be no timber trade and no rice planting, and without these, there would be no Darien. The analogy holds good in still another respect. From time immemorial, the stately African river, at certain seasons, rises and overflows its banks, and so does the Altamaha, as we all know, but just here the analogy ends, for while the presiding deity of the former stream, duly regarding his venerable reputation, regulates the movements of the waters under his control, with tide-like precision each recurring year, the Jolly God of the Altamaha, like other youths of the period has been lately cutting such antics to the discomfiture of rice planters and timber cutters as to arouse a grave suspicion that he exacts a toll of Darien whiskey from the bottle of every raftsman who ventures to return home by the *Daisy*..."⁶

The 1870s and 1880s were the palmy days of Darien's rapid growth as a timber port. A few energetic citizens returning from the war, aided by an infusion of northern capital, had orchestrated the resurrection of the town from the ashes of the war. By the

Timber inspectors at log boom, Cathead Creek, Darien, ca. 1895.

mid-1870s Darien was, according to the civic-minded Grubb, "once again the most important lumber port in the Southern States."[7] In 1878, the Lower Bluff sawmill, the largest in Darien, was acquired by the Hilton Timber & Lumber Company, which soon became a worthy rival to the large Dodge-Meigs interests on nearby St. Simons Island. Joseph Hilton's operation expanded to include mills at Doboy and Union islands near Darien and on Cathead Creek just west of town.[8] During the fall and winter timbering season rafts of upcountry pine logs were floated down the Altamaha to the public booms at Darien where the timber was graded and measured by city-appointed inspectors, then sold to buyers representing the various local timber interests. The booms extended for about three miles along the Darien River on both sides of the town. Legare, in his journal, makes numerous references to the log booms which, in some cases, abutted his rice fields on either side of Generals Island across from Darien.

With the growth of timber came flush times in the town of Darien. In 1889 the Darien Bank was chartered, largely with the influence of the local timber market in mind, and by 1892, had a capital of $50,000. Darien's timber trade thrived in the northeast, particularly in New England, where many of the shipments of lumber were consigned. But with great rapidity in the 1870s and 1880s, Darien's timber trade began to take on an international flavor as the preponderance of shipments began to be marked for destinations in South America, Scandinavia and other parts of Europe and the Mediterranean. Such was Darien's status as an international timber market that the town, with a population which never exceeded 2,000, had four foreign consulates.

In 1888, the Hilton Timber & Lumber Company merged with the Dodge-Meigs firm of New York and St. Simons Island to become the Hilton-Dodge Lumber Company. Headquartered in Darien with Joseph Hilton as its president, this huge corporation dominated the Atlantic coast pine timber market for the next two decades. In 1892, Hilton-Dodge had a corporate office in New York City and was listed with capital in the amount of $1 million, a staggering sum in those days for any commercial concern in tidewater Georgia.[9] "The effort was made and in a short time we found their exertion being crowned with success," Grubb mused when Darien's rebirth was already well under way. "The limits of the city extend[ed] out in all directions fully one mile, with new residences and store houses reflecting credit upon those who had, when they began, no fortunes but their energies and manhood. The mercantile and commercial interests of the town increased...her harbor dotted with flags and ships of other nations, here for cargoes of that, which in days yet to come, is destined to make Darien the lumber port of the world..."[10]

Thomas Hilton, the grandson of Joseph, noted that the 1890s were the "High Water" mark for all this activity, the time that "the flow of the golden-hearted timber down the Altamaha reached its peak."[11] A large sawmill had been built on Doboy Sound, convenient to ocean-going ships entering the harbor. Schooners engaged in the coasting trade arrived from New England and New York to load with Georgia pine timber and lumber to fuel the building boom in the northeast. Foreign ships, usually three-and-four-masted barkentines and square-riggers, used loading grounds in Doboy and Sapelo sounds with as many as eighty-seven vessels being counted in port at one

Lower Bluff sawmill, Hilton Timber and Lumber Company, Darien, ca. 1880.

time loading with Darien timber.[12] By 1900, steamships had begun to take over the trade. Gangs of black stevedores, men of brawn and skill, handled the jobs of loading timber and lumber aboard the ships. Specially-trained blacks ran some of the local sawmills. There were employment opportunities for everyone who wanted work— around the booms, at the loading grounds and in the mills. The work demanded both strength and savvy:

> "The timber was brought alongside ship in the water. Ports were removed from the wooden sailing ships, the long, heavy timbers, muddy from the river, taken in through the ports [in the bow], laid on rollers, rolled to the part of the ship wanted, turned by cant-hooks into place...Each gang of stevedores had a leader who sang a line of a chantey song, the others coming in on the short chorus, then with a 'ho' all heaved together on the big stick in perfect time...Most of the square-riggers came direct to Doboy or Sapelo [sounds] and when they did could bring no cargo, so loaded rock ballast. The loading grounds were selected at the deepest water points...and the ballast rock was dumped on the bank or in the marsh. Large ballast piles are on Union Island and on the Ridge River [at Doboy]..."[13]

After 1900, Darien's economic boom suddenly became a rapid downward spiral due to the overcutting of the upriver forests, which resulted in a steadily diminishing supply of timber. Annual timber export figures reported at the end of each season in the pages of the *Darien Gazette* reflect the decline. From an all-time high of 112.5 million linear board feet of yellow pitch pine timber shipped in 1900, the amount shipped only a decade later was down to a mere 16.8 million board feet.[14] In 1916, the once-dominant Hilton-Dodge Lumber Company went into receivership and the boom times which had revolved around Darien for half a century became little more than a memory. Legare viewed timber developments in McIntosh County with more than passing interest. He realized, as did any astute businessman of Darien, that timber was the factor on which the local economy pivoted. As long as the supply of timber coming downriver each year remained steady, Darien's success was assured. In the heady seasons of the

Timber ships and towboat in Doboy Sound (Sapelo Island in background).

eighties and nineties, the timber flow seemed endless—no one dreamed it would ever run out, least of all Richard Grubb who, even after the decline had set in during the first decade of the 1900s, took a long time in accepting the reality that the end of the era had come.[15]

The local dependence on timber notwithstanding, it was rice cultivation around which Legare and a small number of his McIntosh and Glynn County contemporaries based their livelihood. Rice was the reason Legare had come to Darien in the first place, and it was rice which would occupy all of his professional concerns for more than twenty-five years. It was understood, and accepted, by the handful of local planters that the rice business, once the dominant economic force in the region, was now playing second fiddle in the Darien commercial picture. Rice had become the stepchild to timber locally and there were no indications prior to 1900 that this balance would ever change.

Compared to the peak production years of the 1850s, there were few Altamaha delta planters left during the postbellum period when the rice kingdom was in its twilight. Frances Butler Leigh, daughter of Pierce Mease Butler and Fanny Kemble Butler, had tried with rapidly diminishing profits, to revive her father's Butler's Island rice holdings in the 1870s. In addition to Legare, who planted Generals Island, and later, Champneys, other postbellum planters were A. S. Barnwell, who planted Champneys before Legare, William C. Wylly at Broughton Island, the Strain brothers, Robert and William, who leased Butler's Island from Mrs. Leigh in the 1890s, the three Cathead Creek planters, D. S. Sinclair, C. O. S. Mallard and Thomas Hart Gignilliat, and several on the Glynn County side of the Altamaha delta, including Richard Corbin, Jr. at Hopeton-Altama, and James Troup Dent at Hofwyl.

Legare had acquired his earliest experience in rice planting not long after the Civil War when he worked for a short time in the industry near his Adams Run home in South Carolina. In 1878, soon after arriving in the Altamaha delta, Legare assumed the management of the Generals Island rice plantation, across the north branch of the Altamaha River from Darien, in the employ of his family friend, Nathaniel H. Barnwell, like Legare a native South Carolinian. Barnwell, at the time, was leasing a portion of Generals Island from the island's owners, Frances Butler Leigh and Sarah Butler Wister, daughters of the late Pierce Mease Butler.[16] In January 1896, Legare took on additional duties when he assumed the management of the Champneys Island rice plantation, in the middle Altamaha delta just south of Butler's Island. At Champneys, Legare was in the employ of the Central Railroad Banking Company of Savannah, which had acquired the island on foreclosure several years before.[17] Here, Legare was responsible to the Central Railroad Bank's agent, T. M. Cunningham. Legare remained in this position until retiring from the rice business in early 1905 when the Central Railroad Bank decided to sell Champneys after several years of declining profits. Legare's Journal consistently reflects his affinity for the rice culture of tidewater Georgia. Indeed, the salient feature of Legare's Journal (at least through 1904), aside from documenting family events and matters relevant to Darien society, is his attention to reporting the details of his rice operations, his almost obsessive intent to record the minutiae of day-to-day work at Generals and Champneys islands. Legare was certainly in the right place for a heavy commitment to rice.

The rich, marshy delta comprising the Altamaha River's bottomlands below Darien was highly conducive for the cultivation of rice. Nature had been kind to the region around Darien in this respect. From about 1820 until the Civil War, the Altamaha River valley was the scene of heavy agricultural activity as all the delta islands were cleared and diked for rice cultivation. According to the U.S. agricultural census of 1860 there were under cultivation in this district (McIntosh County) 5,800 acres of rice, yielding 195,000 bushels. This acreage included 1,000 acres on Butler's Island, 1,000 acres on Cambers Island, 500 acres on Broughton Island, 600 acres on Champneys Island, 300 acres on Generals Island and smaller tracts along Cathead Creek west of Darien. The flow of fresh, upriver, water headed seaward on the ebbing tide, was used to irrigate these fields, and tidal-powered steam mills were constructed on the larger plantations to mill the rice. Vestiges of many of the dikes, canals and mill structures are still in evidence throughout the delta.

During the antebellum period tidewater Georgia was known as the "Rice Kingdom" for good reason. During the 1840s and 50s, Georgia was second only to neighboring South Carolina on the world market for the production of rice. Production in the rice districts of the Savannah, Ogeechee and Altamaha rivers reached its peak in 1861, on the eve of the Civil War. The 1860 agricultural census reveals that in 1859 Georgia rice planters produced 52.5 million pounds of rice, compared to 119 million pounds produced in South Carolina. In Georgia, Chatham County (Savannah River) accounted for 25.9 million pounds of rice produced in 1859, followed by the Altamaha River delta (McIntosh and Glynn counties) with 11.2 million pounds, and the Satilla River (Camden County) with 10.3 million pounds. There were ninety-six rice planters on the Georgia tidewater in 1860, twelve of them in McIntosh County and six in neighboring Glynn. By 1870, these numbers were down to eight in McIntosh and four in Glynn. In 1880, the local rice acreage was down from the high of 5,800 acres under cultivation in 1860 to 4,035 acres; in 1900, the acreage was only 1,065, and by 1930 there was no acreage being devoted to the commercial cultivation of rice in McIntosh County.[18] In the 1920s some of the former rice fields on Butler's and Champneys islands were developed experimentally for the production of truck crops.

9

Rice plantations represented enormous investments in capital for the select few planters financially secure enough to risk everything they had on these elaborate operations. Huge outlays in cash were invested in slave labor to work the plantations, in addition to the equipment necessary to construct and maintain the elaborate system of dikes, canals and trunks. Because of slavery, the production of rice proved extremely profitable for the planter and his agents. The largest groupings of the tidewater slave population were those which worked the rice plantations, which were always considerably more labor intensive than the Sea Island cotton and sugar plantations on the barrier islands. Major Pierce Butler, followed by his grandsons John and Pierce Mease Butler, managed upwards of a thousand slaves on their Georgia plantations, more than 500 of which were at the Butler's Island rice plantation; Phineas Miller Nightingale had 169 slaves at nearby Cambers Island and Jacob Barrett worked 155 at Champneys Island.[19]

Certainly the conditions in the lower Altamaha region were highly conducive for the cultivation of rice. Altamaha clay soil is flushed downstream from the uplands of Georgia and is characterized as being the most fertile and strongest soil in McIntosh County.[20] The delta is subject to frequent overflow both by flood waters from upriver and by spring tides, except in areas protected by rice dikes. This fertile area supported rice culture due to its nearness to the Atlantic Ocean, which provided for the irrigation of fields with tidal flow without the serious threat of salt water intrusion. This natural system of irrigation along the coastal rivers of South Carolina and Georgia allowed a narrow strip of land several miles inland from the sea suitable for growing rice to be alternately flooded with fresh water, then drained, on the ebbing tide. The canals and embankments associated with rice development along the river tide swamps enabled the planter to regularly flood his crop in order to kill insects, weeds and grasses, as well as supply the necessary soil moisture. The Altamaha delta soils were consistently renewable as fresh deposits of topsoil were brought downriver from the uplands.[21]

A rice plantation (both before and after the Civil War) was a natural hydraulic system, dependent on the ebb and flow of the river with a complex arrangement of dikes, irrigation ditches and trunks, all designed to function at the behest of the tides. Common practice dictated that diked embankments along the riverbanks be built up about two feet above the maximum high tide line; just above the low tide level, trunks, or floodgates, were built in the embankment to allow water access from the river to the enclosed rice fields. These square, wooden tidegates were critical to the efficient management of the rice field.[22] Fields were flooded as the tidal waters pushed against the tidegates, with the door closing on the ebb tide. The reverse procedure was used to drain the field. Rice fields were laid out and divided into squares of eighteen to twenty acres each and connected by canals six to eight feet wide and embankments extending from the main canal leading to the river.[23] Shallow-draft boats, known locally as "flats" and poled along by workers, were utilized in the canals and ditches of the fields to transport harvested rice to the threshing mill or to provide access to the dikes and trunks to make necessary repairs.

Loose grains of rice which had fallen during the harvest and allowed to sprout and then turned with the stubble of the previous crop, usually matured and became "volunteer rice" the following season. Volunteer rice devalued the new crop and some planters elected to burn off the stubble before turning over their land.[24] Caterpillars also proved troublesome. "The caterpillars are so bad in #9 & 16 that I flow these squares today," Legare reports on May 29, 1899, and adds, "I have never seen one fourth as much volunteer in a crop of rice in all my experience. Area #1 from the mill downward has little in now, but it is thick every where else on the place."

Due to the nature of the soil, the rice fields were broken up by hoeing rather than plowing. The subsoil was comprised of wet muck caused by poor drainage and was harmful to the proper cultivation of rice. After the Civil War, without the benefit of forced slave labor, many rice planters in the Altamaha region utilized draft animals, particularly mules, to do much of the heavy work of the plantation.[25] Legare, in his Journal,

McIntosh County women pounding rice, early 1900s.

often refers to his mules as being important assets to his rice operations on Generals and Champneys islands. "I commence the year with a disaster," Legare notes on the 1st of January, 1895. "My best mule Lou got in a ditch on Generals Island last night and froze to death before she was found this morning."

Preparatory to planting, the fields were burned off to remove the stubble and plantings from the previous year's crop. Tidewater rice planters along the south Atlantic seaboard typically sowed their rice in prepared trenches in order to achieve a more uniform distribution. Plantings generally were spread over four to five week periods, occurring from the middle of March to the middle of June, generally with a sowing rate of two to three bushels per acre. As he notes in his Journal, Legare favored the so-called covered method, which called for the immediate inundation of the rice fields as soon as the seed was covered by the workers. This first flow, called the "sprout" flow, covered the grain with twelve to eighteen inches of water and protected it from birds and insured rapid germination of the seed.[26] Over the next several days, the rice would sprout, after which the water was drained from the field. When the field was completely dry, the rice would take root, followed by a second irrigation called the "point" flow, which served to eliminate weeds which had developed during the dry growth period. The water was again drained, followed by a hoeing then a three-week long "stretch" flow. Then followed a dry growth period of thirty to forty days and a final "harvest" flow lasting about fifty days prior to final draining and harvest. One of the most difficult problems a rice planter encountered in the protection of his crop, in addition to volunteer rice, was that of the innumerable birds that were attracted to the fields to eat the grain. Workers (and children) were employed to create noise and movement to discourage the birds, often to little avail.[27]

Harvest commenced in mid-August and lasted into October with the rice being cut by hand by workers using a sickle, or "rice hook." Dried rice was tied in bundles, or sheaves, and transported on flats floated in the main canals to the mill house for the various stages of processing. These involved threshing to remove the rice from the plant, winnowing to separate the grain from the chaff, and pounding in the mill with heavy pestles to remove the hulls from the rice kernel.[28] Planters who were unable to afford expensive pounding machinery usually sent their rice to market in Savannah as "rough rice" still in the hull. Pounded rice, with the hulls removed, was shipped as "clean rice." Until the time of the Civil War, threshing and winnowing was generally done by hand using flails.[29] Afterwards, threshing machinery became increasingly utilized. Although expensive, Legare refers several times to his threshing equipment as increasing the efficiency of his rice operation and making it worth his investment. Threshing machinery could process up to 1,200 bushels of rice per day. In coastal Georgia, 45 to 55 bushels of rough rice per acre represented a good crop—anything over that was considered excellent. Due probably to the better soil conditions, it was generally acknowledged that the yield per acre for Georgia rice was somewhat higher than that for South Carolina, regardless of the availability or types of harvesting and milling equipment of which a planter might avail himself.[30] The more technology acquired by a planter logically tended toward more efficiency in getting one's harvest to market, but did not necessarily always improve the yield per acre. Such conveniences, however, were not always without their drawbacks—"The mill has done nothing since midday Tuesday when the thresher broke," Legare noted on October 19, 1899. "Two of our flues were leaking also. I have repaired the thresher but find that the flues are in such bad condition that I cannot plug them..."

Legare's occasional problems with machinery notwithstanding, his greatest frustration encountered in a quarter century of rice planting in McIntosh County was undoubtedly that of never quite being able to come to satisfactory terms with his hired labor. As is evident from the numerous references in his Journal, Legare's perceptions of what a day's labor entailed were often at odds with those of his African-American workers. The problem with labor after the Civil War was not unique to Legare. Labor was increasingly expensive—black males often made up to $1.25 a day by the late 1880s— and yet, "the laborers became more and more inefficient, sometimes openly refusing to undertake tasks and in other instances performing the tasks inefficiently."[31] The former slaves and their offspring were almost unanimously resistant to returning to the rice fields with the same degree of labor intensity of the antebellum period, this being the pattern throughout the Georgia-South Carolina rice kingdom. Declining efficiency of labor led to declining profits on the south Atlantic seaboard, while the growing commercial importance of rice cultivation in Louisiana led to that state becoming the leading rice producer by the late 1880s.[32]

The black rice field workers were not unaware of their newly achieved autonomy. The antebellum planter aristocracy of the rice coast was gone forever. "Under slavery the planter resorted to coercion and the task definition of labor. Emancipation eliminated the legal possibilities of coercion. Thus planters found themselves forced to negotiate for labor supplies," Thomas F. Armstrong notes.[33] This inescapable fact was tacitly understood by planter and rice laborer alike throughout the postbellum period. Black workers began to acquire land and were able to supplement earnings by growing crops of their own to sell—corn, sweet potatoes, rice, oats and small amounts of cotton. In the case of Darien, there were the ever-present, and ever-active, sawmills, booms and loading grounds of the thriving timber industry which usually offered wages considerably more lucrative than those offered for toil in a hot, wet rice field. "I find it impossible to get hands," Legare notes in his Journal on October 18, 1898. "The lumber people are paying common laborers from $1.25 to $2.50 per day & rations and the negroes will not therefore work for me for 75 cents per day and so not much has been accomplished so far." Observations such as these are a common refrain throughout Legare's Journal.[34] When Legare began planting under contract on Champneys Island in 1896, one of the incentives he was able to offer potential workers was that of housing at the plantation. Thus, in exchange for small houses and garden plots for their own crops, Legare was able to retain at least a core force of workers the year around.

In essence, the rice industry of the Georgia tidewater, once so dependent on slavery for its ability to generate profits, was never able to recover from the effects of civil war and its attendant loss of a ready (if unwilling) and cost-effective labor force. In addition, raids by Union naval forces had exacted a severe toll on the local rice planters, with the destruction of mills, machinery, tidegates, embankments and other accouterments. These losses were, in many instances, irreplaceable due to the evaporation of working capital held by the former planter aristocracy (what was left of it) as a result of the war. With lack of adequate capital the old system could not survive, and profits were minimal once the former slaves began working for wages.[35] Blacks even began to lease rice lands for themselves, forming cooperative ventures at Darien to pursue their own ventures. Legare makes several references in his Journal to blacks leasing rice fields on Generals Island across the river from Darien. While the majority of local blacks were engaged in the timber, lumber and naval stores industry, those who weren't were almost exclusively involved with agriculture, chiefly the cultivation of rice. Lillian Fox Sinclair, a Darien resident during the postbellum period, recalled that "there was rice planting on Butler's, Champneys, Broughton, Generals and Rhetts islands, which gave employment to hundreds of negro men, women and children. [The children] were bird minders in the fall when the rice was in what they called the milk. The rice birds would come by the millions to suck out the soft grains. The children would yell and clap small pieces of plank together which would frighten them away for a time. These birds one could buy for twenty-five cents a dozen, and were delicious morsels..." With the costs of replacing equipment and rebuilding the plantations, it was the decade of the 1890s before the planters began to rid themselves of the debts they had accumulated from the end of the war. By then, it was only a matter of time before the industry expired altogether, both in Georgia and in South Carolina.[36]

There were other factors besides labor and capital which debilitated the Altamaha rice industry. The increase of upland cotton production in the postbellum period took a heavy toll on the quality of the land and led to decreased fertility in the soils which were carried downstream to the delta. Legare makes reference to a series of freshets during the 1890s which caused severe damage to the delta's agricultural operations as the river overflowed its banks on numerous occasions, bursting tidegates and allowing salt water intrusion in the fragile rice fields. Several hurricanes and tropical storms in the late 1890s were little short of devastating to the struggling rice planters but, by then, rice planting in tidewater Georgia had become almost an anachronism. Legare writes of the 1898 hurricane and tidal wave inundating his Champneys Island rice lands and drowning several of his black workers. The final blow to the Georgia industry came in 1900 when a new variety of rice imported from Japan was introduced with profitable results in Louisiana and Texas, driving down prices so much that more and more south Atlantic planters in South Carolina and Georgia left the business.[37]

Legare harvested his final rice crop on Champneys Island in the fall of 1904 and left the industry for good several months later. He hardly seems to lament the fact:

"Sept. 21 [1904]. I have found hands for the rice fields hard to get. So far fully 150 acres of the crop far over ripe before it could, or can, be cut. I estimate our loss from over ripe rice at 1000 bushels.

"Dec. 13. I have advices that the Central [Railroad] people will probably not plant Champneys Island next year, but will offer it for sale.

"Dec. 20. Some one broke into the mill building on Champneys Island last night and stole about 10 bushels of rice. They also stole 6 or 7 of my fowls...

"Jan. 3 [1905]. Champneys Island was sold in Darien to William C. Wylly for $2,000.00, about 1/3 of its value! And so I am, or will be soon, out of that job..."

A few of his neighbors continued to plant for several more years with steadily declining results until, by 1910, only 584 acres of rice were being grown in all of the Altamaha delta where almost 6,000 acres were being cultivated only fifty years before.[38]

Legare's involvement with civic affairs had begun almost from the time he settled in Darien in 1878. He served as an appointed member of the Board of McIntosh County Commissioners from 1901 to 1905 and, in June 1905, was appointed county clerk, serving in that capacity until February 1927. In May 1904, Legare was appointed clerk and treasurer of the City of Darien at a salary of $50 per month. He held this sometimes difficult, and often controversial and thankless, position until 1927 when he retired from civic life. As city clerk, Legare handled much of the day-to-day activity of running the business of the town handling, among other duties, the chairmanship of the Darien Pilots Commission, which regulated the activities of the local bar pilots and established rates of pilotage for shipping entering the local harbor.[39]

Few citizens or public servants were as loyal and supportive of their community as Legare was in regards to Darien. His efforts on behalf of enhancing the business atmosphere in the town are regularly noted in his Journal. Darien was never a large place. The town's population of 1,739 in 1900 was a peak which was followed by a steady decline in residents, which was the trend through World War II.[40] These were the days when most local businessmen envisioned an inexhaustible supply of timber continuing to come to the town from upriver, little realizing in 1900 that the end of the boom period for Darien was fast approaching.

Before that, however, during the 1880s and 1890s, Darien was every bit a part of Henry W. Grady's vision for a "New South." In 1889 the Darien Short Line Railroad was begun by a group of enterprising businessmen with a view to shuttling timber from the interior to the developing deepwater timber loading grounds of Sapelo Sound, twenty miles north of Darien. Wharfage and other loading facilities were constructed on the Front River east of Creighton Island and on the Julianton River near Harris Neck for the accommodation of the larger ocean-going ships which were arriving for local timber and lumber, rather than risk the shallower waters of Doboy Sound nearer Lower Bluff and the other Darien sawmills.[41] Towboats transported drifts of timber from the Darien mills to Sapelo Sound in ever-increasing amounts and it was easy to envision even greater days ahead for the local industry.[42] No one put much stock in the upriver timber supply running out until it was too late.

Legare, like everyone else, was swept up in the euphoria of the local economic high for a considerable time, but he realized sooner than most that the bubble must burst, that the boom must inevitably turn bust. As early as 1910, Legare writes tellingly, "...Now the question is, what are they going to do about it? And, could such a thing have occurred anywhere else than in *poor old, dead Darien?*" (Editor's italics). Delaying the inevitability of Darien's demise was the town's brief flirtation with the railroad. In early 1895, the Darien & Western Railroad finally completed the laying of track to Darien, giving the town its long-awaited first rail link with the outside world. Legare celebrates the arrival of the first train to the Columbus Square depot on January 26, 1895 with an appropriate entry in his journal and, indeed, it was a significant event. Ten years later, Legare has high hopes for the D. & W.'s new spur track to the Lower Bluff sawmill as a means of conveying timber to the gang saws more efficiently than the old rafting method. But the railroad came too late to save Darien from its irrevocable economic decline in the first decade of the 1900s. And when the Georgia Coast & Piedmont Railroad completed its steel bridges and trestles across the Altamaha delta to give Darien its first rail link with neighboring Brunswick to the south in 1914, the new conveyance was more a convenience for citizens—everyone by then realized that the local timber industry was dead and was not likely to recover.

Still, there were diehards: "We claim without fear of contradiction to be the Garden Spot of the South," a Darien promotional flyer gushed in 1915. "The magic touch of energy and thrift will nowhere yield a larger reward."[43] This was at a time when the shipments of timber from McIntosh County had dwindled to almost nothing and county businessmen were looking to promote Darien as an agricultural paradise, even though it had clearly been demonstrated that rice, at least, was not the way to the county's economic salvation. Events beyond anyone's control had seemed to conspire against Darien in the late 1890s—destructive hurricanes in 1896 and 1898 had proved devastating both to the timber and rice industries. An 1887 fire had destroyed a good portion of the upper bluff business section along the Darien River, including the town's leading hotel, which was never rebuilt.

Darien's problems entailed more than just the business downturn brought on by declining profits in timber and rice—there were serious social concerns as well. Racial tension in Darien evolving from an incident in August 1899 created bitter feelings which took years to overcome. A black man, Henry Delegal, was accused of the rape of a white woman, Matilda Hope, in northern McIntosh County. Local authorities, intending to remove Delegal to Savannah for "safekeeping" were blocked in their efforts by a large delegation of black protesters who insisted on Delegal being kept in the jail at Darien—they feared the proposed transfer of Delegal to Savannah was part of an effort to have him lynched by a white mob.[44] The black "insurrectionists" kept a close watch on events at the jail and each time the sheriff attempted to move Delegal hundreds of blacks gathered, called to the scene by the constant ringing of the bell of the nearby First African Baptist Church. The large black population of Darien and McIntosh County was intent on protecting Delegal, which historian W. Fitzhugh Brundage correctly interprets as "the predictable and understandable expression of the heritage and attitudes of blacks in McIntosh County." At the request of Mayor Spalding Kenan, militia troops from Savannah were dispatched by rail to Darien by Governor Allen D. Candler to maintain order. The situation was on the verge of getting completely out of control when a local white, Joseph Townsend, was shot and killed by blacks he was attempting to arrest for inciting the riot in Darien. Colonel Alexander R. Lawton, ranking commander of the militia, is credited with preventing additional bloodshed as he kept the local posse, bent on avenging the death of Townsend, in check. The black leaders of Darien and McIntosh County, which included preachers, the chairman of the Republican party, the Collector of the Port, and the editor of the local black newspaper, the *Darien Spectator*, prevailed upon the leaders of the riot to turn themselves in and order was gradually restored. Henry Delegal was acquitted of the charge of rape, but twenty-three of the alleged insurrectionists were convicted of rioting and received fines ranging from two hundred fifty to one thousand dollars and prison terms of twelve months hard labor. The sons of Delegal involved in the shooting of the white deputy were convicted and sent to the state penitentiary.[45]

Legare, not surprisingly, takes the traditionalist Southern conservative view of these proceedings in his Journal, which report the events of August 1899 in every detail. He expresses outrage at the murder of Joseph Townsend, echoing the sentiments of most of the town's white populace, and feels the overall tone of the incident has stained the reputation of his town. He leaves little doubt as to where he feels the responsibility lies and who should be blamed for these unfortunate developments which he sees as having besmirched the good name of Darien. However, the protest at Darien clearly demonstrates that shows of force by coastal blacks could intimidate whites and curb the tendency during that period of whites lynching blacks with impunity. The action by the McIntosh County blacks was spontaneous, without leadership or organization but, as Fitzhugh Brundage points out, was effective because of unity in the local black community, which was largely invisible to local white authorities at the time. "Local black leaders did not provoke, lead or control the 'Insurrection.' They only played a role in ending the confrontation and in regaining the trust of local whites," Brundage notes.[46]

Georgia tidewater blacks had, since Reconstruction, gained their strength through sheer numbers as well as their collective sense of unity and purpose. The freed slaves of the coastal rice and Sea Island cotton plantations were in the majority in all of the coastal counties. Legare and his white neighbors failed to understand that in McIntosh County, opportunities for blacks in the growing timber and lumber industry beginning in the early 1870s, along with the political power base established by Tunis Campbell, a black carpetbagger, and community unity of local blacks, had laid the solid foundation for collective challenge and confrontation when they felt threatened. Henry Delegal was much more than a lone accused rapist—he was symbolic of the mounting frustrations felt by the blacks of Darien and McIntosh County.

Despite the occasional periods of unpleasantness, Legare never lost his faith in Darien or the possibilities for its future, even though, for a number of years—even at the peak of the timber boom—he was one of the few who realized the town's glory days were fading. In the winter of 1907, Legare was appointed by the stockholders of the Darien Ice Manufacturing Company as the Director of that establishment at a salary of $130.00 per year. By 1911, through no fault of his own, the business was waging a losing battle for survival. Legare and his son, Houstoun Legare, tried, with rapidly diminishing success, to operate the ice-making plant at a profit, but were ultimately overwhelmed by repeated, and costly, equipment failures.[47] Meanwhile, Darien's attempt to replace the dying timber industry with a shipbuilding venture in 1918 was short-lived. It received the full and enthusiastic endorsement of Legare, as noted in his Journal, with hope for the town's future invisibly written between every line. But Darien would never regain the commercial dominance it enjoyed in the late 1800s. Legare sounds almost despairing in his entry of November 11, 1919: "The Georgia Coast & Piedmont Railroad has been sold to junk dealers, no trains are being run—poor old Darien! Within the last month about five families have left the community, and others must follow if the [rail]road is discontinued. The town is *dead*. The only question to be determined is how deep it will be buried." It seems appropriate to note here that Legare's death in 1932 coincided with Darien's lowest economic ebb.

His civic frustrations aside, Legare's complete and selfless involvement with the Darien Presbyterian Church was his greatest source of refuge. Legare came from a family of devout Presbyterians in South Carolina, and he and his wife, Charlotte, immediately became affiliated with the Presbyterian church in Darien upon their taking up residence in the town in 1878. Legare worked tirelessly on behalf of the church throughout his fifty-five years in the community. For fifty years, right up to his death in 1932, Legare served on the church's Session as a Ruling Elder. No one in the history of the church, before or since, served longer. For thirty-nine of those years he was Sessional clerk. As he grew older he placed increasing faith in his religion, as amplified on a number of occasions in his Journal in which some of his entries unabashedly tend to sermonize. In addition to maintaining the records of the church, Legare researched and wrote a history of the Darien Presbyterian Church in 1900, which served as the primary source of the church's long heritage for most of the twentieth century.[48]

Many of Legare's closest friends were involved with the Presbyterian church—he was particularly close to pastors such as N. Keff Smith and Thomas M. Hunter and he had long and pleasant associations with fellow elders such as Reide Walker (also his Ashantilly neighbor), A. E. Dimmock and Charles O. Screven Mallard.

Legare was fond of taking the civic high road in Darien throughout his professional career as the city clerk. Although constantly answerable to the occasional vagaries of the aldermanic council of Darien, he nonetheless managed the town's affairs with fiscal responsibility and a conscious devotion to his proper duty. Legare enjoyed this role and felt, rightfully so, that he had both earned and deserved the respect of his fellow citizens for his efforts on behalf of the town. Legare's sense of place and permanence are not out of character for that generation of southerners. Among southerners, permanence was equated with one's status of acceptance and respect in a community, particularly in smaller locales such as Darien. In 1882, Legare entered into an agreement with William H. Ingram of Darien by which he purchased, for one hundred dollars, two small lots at Ashantilly near the marsh overlooking Black Island. Being an excellent carpenter, Legare immediately began work on his new home. Later, he moved his former residence, a frame house, from the rice plantation on General's Island to Ashantilly to serve as an addition to the existing structure (a laborious undertaking involving the considerable difficulty of transporting lumber through the tidal creeks on flats). Legare and Charlotte lived at Ashantilly until his death in 1932, as did their children Claudia, Margaret, Emma and Houstoun, and his stepson, Arthur Hamilton.[49]

Legare's input (and insight) is evident in his role in developing a new charter and revised set of city ordinances for Darien, officially adopted in 1910. The new city codes were to remain in effect for decades, a testimony to the wisdom and perspicacity of Legare who seems to have played a major role in compiling the document. The code properly reflected the times of Darien in that it included sections and provisions for the city's board of health, infectious and contagious diseases, port quarantine, cemeteries, rice culture within the city limits, city marshal's sales, timber measurers and inspectors, inspection and measurement of cypress timber, impounding of stock, meat markets, a wild olive ordinance, sale of wood in the city, public drunkenness and closing of barber shops on Sabbath days, among other interesting items. The logic behind many of these ordinances are frequently reflected in the observations made by Legare in his Journal. According to the city code, it was "not lawful for any laborer, tradesman, butcher, artisan or other person to perform or exercise any labor, trade or business on the 'Lord's Day' nor to open any shop, store or other place of business in the City of Darien on the 'Lord's Day....'" Legare personally felt very strongly about this, again as often reflected in the pages of his Journal. No wagoner, or drayman, was permitted to "pursue his usual occupation on the Lord's Day (except to transport baggage of and for passengers to the Railroad or Passenger Steamboat)." There was to be no card playing on Sunday. Horses, mules or other animals could not be hitched to any tree or fence of any citizen without proper consent; it was "not lawful for any junk dealer to purchase any junk from any person before sunrise or after sunset," no bicycles could be ridden on city sidewalks, nor could a person "ride a bicycle on the streets without having a lamp lighted, and at every public crossing he shall ring his bell as a notice to pedestrians." Furthermore, it was unlawful for persons to engage in fighting on the public streets or sidewalks and it was unlawful "for any person to pen and keep up any hog or hogs in the City of Darien at any time of the year," and it was unlawful for any person "in sight of any public street, alley, sidewalk or square of the City of Darien, to bathe in any river or creek or otherwise make an indecent exposure of his or her person in any public place." Additionally, it was not lawful for any railroad train or engine "to cross the Shell Road in the City of Darien without first blowing the usual road-crossing signal and ringing its bell," and it was not lawful for anyone "to sell any raft of timber which comes to the City of Darien without first having it measured and inspected by one of the Public Inspectors of the City."[50]

It seems evident that Legare's piety represents the common denominator between his devotion to the work of the Darien Presbyterian Church and his more secular duties relating to the conduct of the affairs of the City of Darien. It is hardly surprising, therefore, that Legare thoroughly disapproved of the saloons, bars and "houses of ill repute" which made Darien a favorite with rowdy timber rafthands and waterfront stevedores. Throughout his Journal, Legare reflects on the pitfalls caused by the consumption of spirituous liquors. Legare records the obituaries of numerous friends, neighbors and business associates, with appropriate comments reflecting on their impact to the community, and the cause of their demise. "He drank himself to death" is a regular, almost routine, refrain by Legare in noting the deaths of various individuals.[51]

Legare, who took great delight in the growth and progress of his children and immensely enjoyed the company of friends, as well as recreational pursuits such as fishing and bicycle riding, had a seemingly limitless enthusiasm for life, as is consistently apparent in the pages of the Journal. Paradoxically, he displayed—as expressed with almost maudlin regularity throughout his Journal—what seemed to be a morbid fascination with death. Certainly, he had known his share of personal tragedy. He and Charlotte had lost their infant son shortly after its birth in June 1877, an event which profoundly affected Legare. In July 1900, the Journal shows that Legare and Charlotte are devastated by the murder of thirty-one-year old Arthur Hamilton, Legare's stepson, during a labor dispute involving local lumber men and stevedores. Legare's Journal is filled with the accounts of death and his interest in medical matters. He frequently notes his collection of medical books, and the fact that he regularly refers to them, and it is hardly surprising that one of his closest friends for many years is the local physician, Dr. P. S. Clark. The unusual circumstances which surrounded the death of several members of the community merit unusually long entries in the Journal and one can almost visualize Legare diligently consulting his medical books in an attempt to frame his own interpretations of a particularly interesting medical case. But throughout the Journal, and particularly from about 1915 onwards, Legare demonstrates a clear awareness of his own mortality—he never expresses fear or trepidation at the prospects of his own death. One rather feels that Legare has long before made his peace with God and anticipates the death experience with the relish and anticipation that he will presently embark on the great adventure on which so many of his friends and acquaintances have preceded him. "My old friends seem to be dropping off like the autumn leaves," Legare notes on October 31, 1919 in reporting the death of another of his local acquaintances. Yet Legare was to carry on for another thirteen years, and survived his wife by ten years.

Legare died in his eightieth year on the 14th of October, 1932 after a brief illness. He had faithfully kept his Journal almost to the end, recording the events of family and community life as he saw them, his hand growing ever shakier, his handwriting declining almost to the point of illegibility. His daughter, Emma Legare Whitesides, in an addendum to his Journal, recorded his obituary just the way Legare himself would probably have written it: "[Father] was taken sick on Oct. 8th but not seriously enough to cause alarm...but on Thursday he developed pneumonia and became rapidly worse...in spite of the splendid attention he received, on Friday night at 9.15 he died. Today, Sunday, Oct. 16th, he was laid to rest...In spite of a terribly stormy day with rain coming down, nearly every family in the community was represented at the funeral..."

For such an interesting man living in such interesting times, Legare would have been pleased.

BUDDY SULLIVAN

NOTES

1. The literature relevant to the growth and development of tidewater Georgia's rice industry is extensive. The best accounts are Albert V. House, ed., *Planter Management and Capitalism in Antebellum Georgia: The Journal of Hugh Fraser Grant, Rice Grower* (New York: Columbia University Press, 1954); James M. Clifton, ed., *Life and Labor on Argyle Island: Letters and Documents of a Savannah River Plantation, 1833-1867* (Savannah, Ga.: Beehive Press, 1978); Julia F. Smith, *Slavery and Rice Culture in Low Country Georgia, 1750-1860* (Knoxville: University of Tennessee Press, 1985); William Dusinberre, *Them Dark Days: Slavery in the American Rice Swamps* (New York: Oxford University Press, 1996); and James M. Clifton, "Hopeton: Model Plantation of the Antebellum South," *Georgia Historical Quarterly* 66 (Winter 1982), 429-49. For the economics of rice culture the outstanding documented study is Peter A. Coclanis, *The Shadow of a Dream: Economic Life and Death in the South Carolina Low Country, 1670-1920* (New York: Oxford University Press, 1989).

2. For a detailed general review of Altamaha delta rice plantations, particularly those in the McIntosh County portion of the delta near Darien, see Buddy Sullivan, *Early Days on the Georgia Tidewater, The Story of McIntosh County and Sapelo* (Darien, Ga.: McIntosh County Board of Commissioners, 5th edit., 1997).

3. Legare writes of these experiences in his introductory memoir compiled in 1900 to preface the new journal he began after his earlier journal was damaged in the 1898 hurricane.

4. Detailed accounts of the burning of Darien may be found in Sullivan, *Early Days on the Georgia Tidewater*, 292-309; Spencer B. King, *Darien: The Death and Rebirth of a Southern Town* (Macon, Ga.: Mercer University Press, 1981); Luis F. Emilio, *Brave Black Regiment: History of the Fifty-Fourth Regiment of Massachusetts Volunteer Infantry, 1863-1865* (Boston: Boston Book Co., 1891); Russell Duncan, ed., *Blue-Eyed Child of Fortune: The Civil War Letters of Colonel Robert Gould Shaw* (Athens: University of Georgia Press, 1992), 43-45, 342-46.

5. Delma E. Presley, ed., *Dr. Bullie's Notes: Reminiscences of Early Georgia and of Philadelphia and New Haven in the 1800s by James Holmes, M.D.* (Atlanta: Cherokee Publishing Co., 1976),

xx. For insights into editor Richard Grubb and the early years of the *Darien Timber Gazette*, see Sullivan, *Early Days on the Georgia Tidewater*, 353-57.

6. *Darien Timber Gazette*, May 30, 1874.

7. *Ibid.*, July 9, 1875.

8. Sullivan, *Early Days on the Georgia Tidewater*, 439-50. An interesting, largely anecdotal, account of the Hilton family's involvement in Darien's timber industry is contained in Thomas Hilton, *High Water on the Bar* (Savannah, Ga.: privately printed, 1951). See also Hilton Family Papers, Collection 387, Georgia Historical Society, Savannah: Box 1, Folder 5 contains the Thomas Hilton memoir and other papers relating to the Hilton-Dodge Lumber Company.

9. *Darien Timber Gazette* and *Darien Gazette*, Annual Timber Reports, 1888-1914; Hilton Papers, Coll. 387, Georgia Historical Society (cited hereinafter as GHS): Box 1, Folder 3, deeds of Joseph Hilton for lots in Darien and shares of stock in the Hilton-Dodge Lumber Company; Box 2, Folder 12, agreement of sale of St. Simons Mills to Hilton-Dodge Lumber Company, 1 December, 1888; Box 2, Folder 12, Journal of Joseph Hilton, 31 October, 1891 to 30 November 1893.

10. *Darien Timber Gazette*, July 9, 1875; "Howard's Directory of Brunswick, St. Simons, Darien and St. Marys for 1892" (Brunswick, Ga.: 1892). Rosin and turpentine were also shipped in large quantity.

11. Hilton, *High Water on the Bar*, 9.

12. *Ibid.* See also Carlton A. Morrison, "Raftsmen of the Altamaha" (Masters thesis, University of Georgia, 1970); Hilton Papers, Coll. 387, 1:8, GHS, Lillian Sinclair, "My Recollections of Darien in the Seventies and Eighties," typed ms.

13. Hilton, *High Water on the Bar*, 10. Ballast deposits on which small hammocks (islands) have developed in the salt marshes abound in McIntosh County waters, particularly around Doboy Island, Lower Bluff near Darien, the Front River opposite Creighton Island and the Julianton River near the south end of Harris Neck, the latter two areas convenient to shipping entering and clearing Sapelo Sound. See Sullivan, *Early Days on the Georgia Tidewater*, 450-52, for a description of these areas.

14. *Darien Gazette*, Annual Timber Reports, January 19, 1901, November 5, 1910. Darien's timber decline is documented in Sullivan, *Early Days on the Georgia Tidewater*, 535-72, and Morrison, "Raftsmen of the Altamaha," 81-97.

15. John G. Legare Journal and various issues of the *Darien Gazette* from 1900 to 1913. The rotting pilings of waterfront wharfage at Lower Bluff (present site of the Fort King George State Historic Site) and the brick foundation ruins of the adjacent sawmills remain as silent testimony to the timbering days at Darien.

16. In April 1876, James Walker of Darien sold to Nathaniel H. Barnwell 882 acres of cultivable riceland on the lower end of Generals Island between Generals Cut and Threemile Cut. The Butler family portion of Generals Island, which Barnwell leased, was that part of the island west of Generals Cut. A deed dated January 7, 1907 finalized the sale from Frances B. Leigh and Sarah B. Wister to Robert A. Strain, Robert Manson and John D. Clarke, for $1,500, Generals Island west of Generals Cut, 800 acres total, of which 300 acres were ricelands. This was the portion overseen by Legare from 1878 through 1896. Deed Record Book A, 624-27; Book I, 160, Superior Court records, McIntosh County, Georgia; Thomas Porcher Ravenel Papers, Coll. 649, 20:60, GHS, Leigh-Wister papers, 1889.

17. In February 1871, Charles Green, Jr., of Savannah sold to Archibald S. Barnwell of Darien 937 acres of rice tideland comprising Champneys Island, and all "buildings, farm implements, machinery, etc." pertaining to the plantation there, for $23,000. In February 1888, after Barnwell had lost the property through foreclosure, Edward M. Green sold the island to the Central Railroad and Banking Company of Georgia, for $20,000. Legare took over the management of Champneys from A. S. Barnwell in 1896, overseeing the plantation until February 1905 when the Central Railroad Bank sold the island, for $2,000, to William H. Strain of Darien (not William C. Wylly as Legare states in his Journal, entry of January 3, 1905). In 1923, Richard H. Strain sold Champneys Island to Tillinghast L. Huston of New York. Huston also acquired

neighboring Butler's Island during this period, a sale formalized in his name in May 1927. Deed Record Book A, 583; Book D, 190; Book H, 439; Book S, 518; Book T, 196; Book 1, 277, Superior Court records, McIntosh County, Georgia.

18. *Seventh Census of the United States, 1850, Georgia, McIntosh County,* Agricultural Schedules; *Tenth Census of the United States, 1880,* Agricultural Schedules; *Twelfth Census of the United States, 1900,* Agricultural Schedules, Part II, *Crops and Irrigation; Fifteenth Census of the United States, 1930,* Agriculture.

19. *Eighth Census of the United States, 1860, Georgia, McIntosh County,* Slave Schedules; Sullivan, *Early Days on the Georgia Tidewater,* 181.

20. G. L. Fuller, B. H. Hendrickson and J. W. Moon, *Soil Survey of McIntosh County, Georgia,* Series 1929 (Washington, D.C.: U.S. Department of Agriculture, 1930), 22-23.

21. *Ibid.*, 23. See also Lewis Cecil Gray, *History of Agriculture in the Southern United States to 1860* (Washington, D.C.: Carnegie Institution of Washington, 1933), 721. This account, like House's *Planter Management*, while focusing on the antebellum period, is nonetheless highly useful for its general observations regarding the technicalities of rice culture, most of which continued to apply in the postbellum era of rice planting.

22. House, *Planter Management and Capitalism in Antebellum Georgia,* 24-26; Manigault Papers, Coll. 1290, GHS.

23. Victoria Reeves Gunn, "Hofwyl Plantation" (Unpub. ms., Georgia Department of Natural Resources, 1975), 11.

24. *Ibid.*, 13; Manigault Papers, Coll. 1290, GHS.

25. House, *Planter Management and Capitalism in Antebellum Georgia,* 27; Hofwyl Plantation museum and archives, Brunswick, Georgia.

26. *Ibid.*, 28-32; Gray, *History of Agriculture in the Southern United States,* 727; John G. Legare Journal.

27. House, *Planter Management and Capitalism in Antebellum Georgia,* 34; Gray, *History of Agriculture in the Southern United States,* 729; Buddy Sullivan, ed., "The Agricultural Journal of Roswell King, Jr., 1845-1854" (Unpub. ms., 1997). Portions of John G. Legare's Journal contained in this volume for the period 1899-1904 relate details of the technical side of postbellum rice planting.

28. Gray, *History of Agriculture in the Southern United States,* 729-30; House, *Planter Management and Capitalism in Antebellum Georgia,* 59-65. For methods of cleaning rice and mill machinery, see Keith Read Collection, Coll. 648, 25:287, GHS.

29. Frances Anne Kemble, *Journal of a Residence on a Georgian Plantation in 1838-1839,* edited by John A. Scott (New York: Alfred A. Knopf, 1961), 109, 110; John G. Legare Journal; and Keith Read Collection, Coll. 648, 25:287, GHS.

30. Duncan Clinch Heyward, *Seed From Madagascar* (Chapel Hill: University of North Carolina Press, 1937), 41-42; Gray, *History of Agriculture in the Southern United States,* 730-31. About two hundred bushels yielded one tierce (barrel) of clean rice.

31. James M. Clifton, "Twilight Comes to the Rice Kingdom: Postbellum Rice Culture on the South Atlantic Coast," *Georgia Historical Quarterly* 62 (Summer 1978), 148.

32. *Ibid.*, 151; Stewart Huston Papers, Coll. 1267, 2:17, GHS, contains observations and data on the postbellum rice trade in tidewater Georgia.

33. Thomas F. Armstrong, "From Task Labor to Free Labor: The Transition Along Georgia's Rice Coast, 1820-1880," *Georgia Historical Quarterly* 64 (Winter 1980), 439; John G. Legare Journal.

34. For a contemporary, if somewhat chauvinistic, account of the unsettled labor situation attendant to the revival of the Darien area rice industry during Reconstruction see Frances Butler Leigh, *Ten Years on a Georgia Plantation Since the War* (London: R. Bentley & Son, 1883). See also Russell Duncan, *Freedom's Shore: Tunis Campbell and the Georgia Freedmen* (Athens: University of Georgia Press, 1986) and Sullivan, *Early Days on the Georgia Tidewater,* 332-401, 816-18.

35. Gunn, "Hofwyl Plantation," 77.

36. *Ibid.* Also, John G. Legare Journal, which contains references in increasing frequency in the late 1890s and early 1900s to the declining profitability of rice planting. The quote about children and "rice birds" is from Lillian Fox Sinclair, "My Recollections of Darien in the Seventies and Eighties," typed ms. in Hilton Papers, Coll. 387, 1:5, GHS.

37. *Darien Gazette,* October 8, 1898; Gunn, "Hofwyl Plantation," 77-78.

38. *Thirteenth Census of the United States, 1910,* Agriculture, and *Eighth Census of the U.S., 1860;* Fuller, Hendrickson and Moon, *Soil Survey of McIntosh County,* Series 1929, 8.

39. Minutes of the Board of Pilot Commissioners, City of Darien, 1880-1930, Darien City Archives. See Sullivan, *Early Days on the Georgia Tidewater,* 457-61, for a review of bar pilots and their activities attendant to navigation in McIntosh County during the postbellum period.

40. *Twelfth Census of the United States, 1900,* Georgia, McIntosh County, Population Schedules. See Sullivan, *Early Days on the Georgia Tidewater,* chapters 13 and 14. From the high in 1900, Darien's population had dwindled to 937 by 1930.

41. *Darien Timber Gazette* and *Darien Gazette,* Shipping Arrivals and Departures, various issues, 1889-1912 (the *Timber* in the name of the Darien weekly newspaper was dropped by editor Richard Grubb in 1893). See Sullivan, *Early Days on the Georgia Tidewater,* 488-98, for the development of Sapelo Sound as a deep water timber loading anchorage by the local firms.

42. *Darien Timber Gazette,* September 21, 1889.

43. Darien Board of Trade, Promotional Brochure, 1915. This flyer went on to note that Darien "has an abundance of artesian water, is electrically lighted, [and] has adequate railroad and water communication. The Georgia Coast & Piedmont Railroad crosses the river from the south to Darien, skirts the [McIntosh County] coast and then turns to the west. The Seaboard Air Line Railroad, Savannah to Jacksonville division, is to the west of Darien and bisects the county. Our railroads offer excellent transportation facilities. The Quebec-Miami highway, which enters the county from the north and runs through its entire length to Darien, is but a part of the road system being built with convict labor...Our coast country attracts many people from the leisure class seeking conditions of life which afford health and amusement midst surroundings of beauty. The rigors of the north are unknown here. 'Tis a health giving country...No section presents better agricultural advantages. McIntosh County has a great variety of soils. It possesses both high lands and delta lands, believed to be among the richest on earth...Fairhope [begun in 1912 and later known as Pine Harbor] is a section of our county being developed by Ohio capitalists. A factory [there] is in operation which makes ladders and porch furniture. A large hotel is under construction. Fairhope will speak for itself. Sapelo Island, one of our largest sea islands, is being energetically developed. Its new owner [Howard E. Coffin of Detroit] is carrying on farming and stock raising on a large scale." For the Georgia Coast & Piedmont Railroad's ambitious undertaking to bridge the Altamaha delta with trestle work, see *Darien Gazette,* July 2, 1910, August 3, 1912 and issues ff.

44. *Darien Gazette,* August 26, 1899; *Savannah Morning News,* August 24-27, 1899; John G. Legare Journal.

45. McIntosh County, Georgia, Superior Court Minutes, Book E, 1896-1905, 174-97, 227-28. The most complete, and insightful, account of this incident is contained in W. Fitzhugh Brundage, "The Darien Insurrection of 1899: Black Protest During the Nadir of Race Relations," *Georgia Historical Quarterly* 74 (Summer 1990), 234-53.

46. W. Fitzhugh Brundage, "Black Response to White Violence, 1880-1930," Address to the Georgia Historical Society, May 1992.

47. John G. Legare Journal, April 11, 1907, August 15, 1911; *Darien Gazette,* June 24, 1911.

48. Buddy Sullivan and William G. Haynes, *History of the First Presbyterian Church of Darien, Georgia* (Darien, Ga.: First Presbyterian Church, 1986) and Sullivan, *Early Days on the Georgia Tidewater,* review in detail the history of Legare's church, which was founded by Scottish Highlanders in 1736 making this the earliest Presbyterian congregation in Georgia.

49. Deed Record Book C, 195, Superior Court records, McIntosh County, Georgia. John G. Legare Journal, March 10, 1882, May 1, 1888. The Ashantilly house and property remain in the Legare-Britt family at the time of this writing.

50. *Charter, Revised Code of Ordinances, Rules of Council and Penal Laws of the City of Darien* (Savannah, Ga.: The Morning News, 1910).

51. For example, John G. Legare Journal, January 13, 1890.

Woman fanning rice, McIntosh County, early 1900s.

John Girardeau Legare's Journal

Introductory

I, John Girardeau Legare, believing that an account of the events that have transpired during my lifetime will be of interest to my children and perhaps to others also; I will try to write of them and recount them as they are remembered by me. I have kept a diary for many years in a somewhat imperfect manner. In 1892 I wrote up these memoranda in diary form. But my diary having been carried away in the great tidal wave of October 2nd, 1898, and recovered three months later in a badly damaged condition; I am compelled to re-write it. This I will try to do during the present year—1900.

I was born at Adams Run, St. Pauls Parish, Colleton District (now Colleton County), S.C. on Monday, April 26th, 1852. My father was James Legare, Jr., the son of Francis Yonge Legare, the son of James Legare, Sr....The family is of Norman origin, the name Le gare being Scandinavian, and means "The Owner" or "The Proprietor." The goldsmith Francois [Le gare] was married twice. He and his first wife were Roman Catholics.

He had two sons by his first marriage. My own researches lead me to believe that the elder Francois emigrated to New England. Francois the goldsmith's second wife was a Protestant; she had one son, our fore-parent, Solomon the Huguenot, who emigrated after his father's death to Charleston, S.C. either during the latter part of 1685 or early in 1686...I am the living head of the family of the Huguenot descendants, the descent being as follows—First, Solomon the Huguenot, then his first son Solomon the second who married Miss Mary Stock of Charleston, S.C., 3rd his son Solomon the third who, dying without male heirs, the headship descended to Thomas, 2nd son of Solomon the 2nd. Then James, eldest son of Thomas. Then Francis Yonge, eldest son of James. Then Motte Washington, eldest son of Francis Y. Motte W. dying without male issue, the headship descended to James, Jr., second son of Francis Y., and I am eldest son of James Legare, Jr. My Father, James, Jr., the 10th child of F. Y. & Mary W. Legare, married Claudia Mary Girardeau, daughter of John B. Girardeau of Charleston District, now Berkeley County, S.C., at Adams Run, S.C. January 24th 1850...

Solomon the Huguenot settled in Charleston and Charleston was the home of the family, at least during the summer months for five generations down to grandfather Francis Yonge Legare. Thomas, my 4th ancestor, had his winter home and cotton plantation on Johns Island where a large number of my relatives are buried, including my ancestors Thomas, James, Francis Y. and their wives, and Uncle Motte Legare.

My great-grandfather and my grandfathers home was on the eastern side of the Tugaloo River in St. Pauls Parish about five miles in a direct line from Adams Run, but it is 12 miles by road...My family have always been zealous Presbyterians...

The Girardeaus—Pierre Girardeau emigrated from Tulmont, France to Charleston District, S.C. in 1685 and lived and died there. His son Isaac had two sons, John and Pinckney. Pinckney Girardeau emigrated to Liberty County, Ga. about the year 1735. John, the older son, remained in South Carolina. His son John married Miss Williams. His children were John Bohun, Isaac and a daughter who married a Mr. Limbaker of Georgetown County, S.C. She left two sons, Isaac G. and John G. Limbaker. I suppose their descendants, if not themselves, still live on the South Santee in Georgetown County.

Uncle Isaac Girardeau (grandfather's brother) married a Miss Cox of Charleston. They had one child, Campbell Girardeau, M.D. who was living, and married, in Beaufort County, S.C. when I heard of him last. Grandfather John Bohun Girardeau was twice married. He first married Claudia Hearne Freer of James Island, S.C. Their children were Lafayette, Thomas Jefferson, Emily, Claudia Mary and Edward Freer. Grandmother Girardeau died when Uncle Edward was only five days old of peritonitis and grandfather then married Miss Mary Hughes of Charleston who became the mother of the following—William Hughes, Mary Hughes, Isaac, John Bohun, G. Maurice and Beulah Hughes Girardeau...

Mother, fourth child of John Bohun and Claudia Hearne Girardeau who, strange to say was named for both her mother and step-mother (Claudia Mary), my step-grandmother having been a particular friend of grandmother's. Mother was born in Charleston April 27, 1831...Father and mother were married at Adams Run, S.C. Jan. 24th, 1850. They had 12 children [who] lived all their lives, except during a part of 1865, in Adams Run. Father was appointed postmaster of Adams Run in 1846 and held the office until the time of his death.

All of my Legare kin down to grandfather Francis were accounted rich people. Grandfather must have left an estate of over $100,000, but there were eight heirs-at law and so father's share was somewhere between $12,000 and $15,000. Father was a poor manager and soon lost the most of it, so that I may say that I was always poor. Father had a store in Adams Run and several negroes who he used to hire out, hired themselves out, and so were about like free people...

Wife's relations—Charlotte Marshall Smith was the third daughter and fifth child of Marshall Robert Smith and his wife Elizabeth A. Caldwell Gillam. She first married John Middleton Hamilton of Charleston, S.C. (he was the son of Daniel Heyward Hamilton, M.D. and Rebecca Motte Middleton—both my distant relatives through the Mottes) in 1867. Mr. Hamilton died of consumption Feby. 26th, 1870 and lies buried in the cemetery to the Presbyterian Church at Ninety-Six, Abbeville County, S.C. They had one son, Arthur St. Clair Hamilton, born at Ninety-Six Nov. 26th, 1868...

1860—The War was brewing and these were times of great excitement...About this time two "pine knots," as we called the backwoods people, two men, were at a political meeting in Adams Run, and they were rather the worse for liquor when one of them stepped on the others foot and demanded that the man he had stepped on should apologize to him, the offender, for the injury done, and it amused us all very much.

1861—"Rumors of War." South Carolina had passed an ordinance of secession and everywhere war preparations were being made. A company of infantry was organized in our vicinity...It was named the "St. Pauls Rifles." Nearly all the men of the community joined this company, Father among them. Even the children wore blue ribbon cockades...When Fort Sumter was bombarded April 12th we heard the guns distinctly, although 25 miles away. That event threw everyone into a whirl of excitement. Our local company was sent to guard the mouth of the North Edisto River with some old guns that would look like playthings for children. They were soon after ordered to Virginia and I saw no more of them until the war ended. But Father did not go with them, being Postmaster he was left behind and later he joined the "Rebel Troop," Co. I, 3rd Regiment of S.C. Cavalry. At this time Adams Run was a quite respectable village of 32 dwellings, two churches in the village, two school houses, one for boys and one for girls, an Agricultural Society and Hall and a Library of probably several thousand volumes. There were three ministers, Episcopal, Methodist and Presbyterian, and three practicing physicians, three stores, blacksmith shop, &c.

1864–Dark days! Hard times! No tea, coffee, or any other article. The following are some of the prices that prevailed at this time on account of the blockade of southern ports by the U.S. Navy: calico $1.50 per yard, books $500, shoes $200 per pair, a suit of men's clothes $1500...a silver dollar was worth about $10 in Confederate money...The southern people were an extravagant lot. We were considered poor but we kept five servants to do the work now done by myself, wife and daughters. We had a woman that did nothing but cook, another woman who was wash woman, house woman and nurse and she had a grown woman as assistant. Then we had a man who was butler, milked [cows], attended the horse, got wood &c. with the assistance of a grown man who, when there was nothing else to do, fished and got oysters &c., &c.

1865–On account of Sherman's march from Savannah, Ga. to Columbia, S.C. Adams Run was evacuated on the 16th February. All the country to the south of us had been evacuated; the troops being withdrawn to reinforce Joe Johnston's army. The last train left Adams Run station on the Charleston & Savannah railroad on that date and Father took us to Charleston on the last passenger train, he having obtained a furlough in order to take us to a safe place. I was then a boy of 13 years and quite small for my age. Our train was loaded with sick and wounded soldiers. Trains did not travel as fast then as now. We left Adams Run at about 2 o'clock P.M. and arrived in the car yard of the railroad west of Ashley river about dark having made the trip of 23 miles in about four hours...

Saturday, Feby. 18th, 1865–The most awful day of my life, the experiences of that day will not be forgotten. The sights were grand beyond description, but inexpressibly awful! Boy-like, I was everywhere I could go and saw all that was to be seen; and many of the scenes are as vivid today as they were years ago. Gen. P. G. T. Beauregard commanded the city at that time, and also the Confederate forces in the state of South Carolina, Georgia and Florida and it has always seemed to me that he showed great lack of judgment at the evacuation of Charleston. Of course, I don't know his reasons, but I *do* know the result of his actions and orders. Beauregard evacuated the city early in the morning. Mother awoke us at daylight to take leave of Father who was leaving to join his command, the "Rebel Troop," Co. I, 3rd S.C. Cavalry.

Gen. Beauregard had given orders that all cotton in the city should be burned at daylight. In the yard in which we were staying there was an old mill building stored full of cotton–about 300 bales, then there was another large building just across a canal about 20 feet wide also stored full of cotton. This cotton was ordered burned and was entirely consumed by fire at the appointed time. We knew the day before that the order had been given, and through the intervention of Mr. Edward M. Barnwell, Mayor McBeth sent the fire department to protect our dwelling. But they could not remain because about sunrise there was a terrible explosion over in the northeast portion of the city, which we soon heard was the Northeastern Railroad Depot, which had been blown to pieces, killing and maiming a large number of people and which started a fire that consumed probably 10 or 12 blocks of buildings. This caused the fire department to leave us except for three hand engines manned by colored men. These, however, saved our dwelling.

About sunrise there were a succession of explosions in the Cooper river section of the city (we were near Ashley river). The ironclad gun boats *Palmetto State* and *Chicora* were blown up and I distinctly saw the smoke of one of them...Charleston seemed to be on fire in all directions, huge pieces of lint cotton would float several hundred feet high burning as they went...To my way of thinking the most foolish order given was to destroy the "New Bridge" across Ashley river, just lately finished at a cost of about $100,000. This order was carried out. Then, horrible to even relate, was an order given to blow up the Arsenal which would doubtlessly have killed hundreds of people and have accomplished no useful purpose. We—the children and Mother—sat for probably two hours huddled together expecting death in a most revolting form at any minute for we were quite near the Arsenal. I did not know until years after the war ended why this order had not been executed. Since I came to Darien I met a Mr. Bingley, a traveling salesman who told me all about it. He said that he and others were ordered to execute this diabolical order, but they were to destroy New Bridge first, this work consuming some time, that they then rode to the Arsenal to execute their order, but when they got there they found Mr. Ebenezer Thayer, a man of about 70 years, standing in the gate way with a double-barreled shotgun which he told them was loaded with buckshot, and he told them they would have to enter over his body, that he would shoot the first one that attempted to pass. Mr. Bingley said that Mr. Thayer evidently meant business, and as none of the men seemed anxious to stop a load of buckshot and as they knew the United States troops had already entered the city—coming in from Morris Island—they thought best to leave Mr. Thayer in possession, which they did. The federal troops entered the Arsenal about 11 o'clock a.m. and they probably saved Charleston from destruction by fire started by its friends(?) We expected to be ill treated by the Yankee soldiers but were very agreeably surprised...I certainly hope that the last great day will not have for me or mine such terrors as that day Feby. 18th 1865 had...

1869—I first worked on a rice plantation this year. On Sept. 20th, I went to work as a general assistant for Mr. John D. Parker, a Boston man who planted Col. Morris' plantation at Wilton Bluff. I was engaged for one month only.

In December, Mr. Joseph D. Taylor employed me as a sort of carpenters apprentice at 75 cents a day to assist in building a house for him at Adams Run station. He took great pains to show me how the work was to be done. He therefore did me the kindness of starting me in a way to make an honest living. He also taught me arithmetic. I finished his house the following spring and entered the service of the Savannah & Charleston railroad on the 25th of May, 1870 as a carpenter at $25 per month and rations and I was put to work on a new depot building they were putting up at Adams Run station...

1872—Early in this year Father and I bought the dwelling house in Adams Run, now owned and occupied by my brother Frank. We were to pay $600 for it. Of this amount I paid $375 and Father and Mother paid the balance. I afterward gave my interest to Mother, and Frank wheedled her out of her home and now owns it...

1874—On October 11th I had my church membership transferred to the Presbyterian church at Ninety-Six and took dinner with Dr. N. Hart and family in Ninety-Six when I met Mrs. Hamilton (wife) the second time...Mrs. Hart was something of a matchmaker and she concluded that a match between Mrs. Hamilton & myself was about the right thing and she soon had matters satisfactorily arranged...On Oct. 15th, Bro. Frank and myself went to board with Mrs. Hill—wife's mother—as she lived quite near our sawmill...On Oct. 22nd, Mrs. Hamilton and myself became engaged to be married. On Nov. 24th we were married at her old home, Homestead...

1875—I sold my interest in the sawmill [and] concluded to try planting cotton for a living, I therefore rented the Homestead farm of 65 acres from my mother in law, Mrs. E. A. C. Hill. I planted some corn, oats &c., but mostly cotton; I did not get rich at it that year. Sept. 23rd, 1875 our first child, Claudia Girardeau, was born, at Homestead, Ninety-Six Township, Abbeville County, S.C...

1876—Dec. 27th, 1876 I received a telegram informing me that Father had died the day before of an illness of only about 24 hours—of which I knew nothing—of some bowel trouble, most probably appendicitis. The telegram was a great shock to us. Father was an unsuccessful man. He was too good-natured to say *no*; and consequently he was often imposed upon...For five years before his death he and I were more like brothers than like Father and son. He wanted my opinion and advice on everything he did of any importance. I do not think he had an enemy...

1877

I planted cotton again this year and made such a complete failure that I concluded to quit it...June 14th, 1877 our second child, but first son, James, was born at Homestead. He was a fine, healthy child but took sick when six days old and died of infantile lock-jaw June 23rd, 1877 after suffering agony for about 63 hours. As young as he was I am sure he knew me and would cry for me when I put him down. I held him almost all through his illness, night and day, wife being in bed, and the poor little thing could not suck. We had to keep him under the influence of chloroform...

Early in September I received an offer of business from an old friend, Mr. N. Heyward Barnwell,[1] formerly of Adams Run, but now of Evelyn, Glynn County, Georgia, which I gladly accepted.

Sept. 28. I left Adams Run for Georgia, leaving wife and children with Mother. I arrived in Savannah about midday and as my train over the Savannah, Florida and Western Ry. did not leave until 3 o'clock, I had three hours to spare and spent them in taking in the sights. Savannah was then much smaller and more insignificant than now.[2] There were few houses south of Gaston street, that part of the city being an old field, at least open ground. Still, it was a pretty place.

I left Savannah at 3 o'clock P.M. for Jesup where I took the Macon and Brunswick

R.R. for Stirling Station #1[3] where I arrived at 8 P.M. and had to spend the night. Next morning I started for Evelyn, Mr. Barnwell's plantation six miles away on foot. It was a rainy time and I must have walked a mile of the distance in water, but I did not mind a little thing like that in those days. I was so poor that Mr. Barnwell had to lend me the money to travel on. Soon after I got to Evelyn Mr. Barnwell took sick and had to leave the place, and I a perfect greenhorn at rice planting had to take entire charge of everything. He kept a store in connection with his plantation and I had to keep the books, manage the harvesting and threshing and shipping of the crop, pay the hands and at odd times I did carpenters work. He paid me $40 per month and gave me my board and had my washing done, until my family arrived when, in lieu of board, he paid me $52 per month with no "trimmins" as the darkies say.

December 4. My family[4] arrived at Stirling from Adams Run at about 9 o'clock P.M. in a rain. I met them there in an open buggy and took them to Carterets Point,[5] Mr. Barnwell's summer house and five miles from the rice plantation, Evelyn. We arrived there about 11 o'clock. It was still raining and it was as dark as Egypt, but we were happy.

We stayed there until Jan. 23rd, 1878 when we removed to Darien.

1878

January 23. I am employed by Messrs. A. McC. Duncan and N. H. Barnwell to take charge of Generals Island[6] plantation, just across the river from Darien, Ga., as Overseer at $50 per month, so I left Carterets Point with my family for Darien today. There being no dwelling on the plantation we had to rent a house in Darien. Darien was a different place then from what it is now [1900]. It was in the height of its prime[7] with probably three times as many white people living in it as now and fully three times as much business was done there. We had some difficulty in getting a house to stay in. There were only two vacant houses in town. Neither of these were good ones but we had to take what we could get. So we hired and lived in the little house belonging to Mr. Adam Strain[8] between his dwelling and the Hilton office.[9] We occupied that house 18 months before we could get a better one.

We got to Darien at 8 o'clock at night having come over from Evelyn in a flat and we had a time of it getting our things up from the wharf, putting them to rights and getting sleepy children to bed. Fortunately Mr. Duncan had engaged John Grant, a responsible colored man who, strange to say, had married Jane, a woman who used to belong to step grandmother Girardeau, and he assisted us greatly. We found the people of Darien very friendly, sociable and helpful. Soon after we got to Darien, Arthur upset a kettle of hot water on he and Claudie, burning both about the body quite badly as wife was not well. Our Pastor's wife, Mrs. Henry F. Hoyt, took them in charge and used to dress their wounds. Those we found especially helpful and friendly were Mrs. Adam Strain, Mrs. Hoyt, Mrs. E. M. Blount[10] and Mrs. Churchill.[11]

May 25. Our 3rd child and 2nd daughter, Margaret Hart, was born at 11:30 A.M. Wife was dangerously ill and Dr. R. B. Harris[12] had to use an electric battery to keep her alive. Mrs. Hoyt was with us and assisted greatly. In fact, I don't know how we would have done without her. She continued her ministrations as long as they were needed during my wife's extended illness which lasted several weeks.

Sept. 11. A severe N.E. gale today flowed up Generals Island, and other places also, so deeply that the rice floated, rafting it in several different places on the plantation. We had to put the mules in the mill. The gale was accompanied by heavy rains and great damage was done. Our banks[13] were broken in one place—near #5 trunk. The loss to the crop was estimated to have been about one third. The following should have appeared earlier.—

March 4. I had my membership transferred from Ninety Six to the Darien Presbyterian Church. The Rev. Henry F. Hoyt[14] was pastor and the following were Elders: Theophilus P. Pease,[15] R. K. Walker,[16] T. Shep. Quarterman,[17] and L. B. Davis.[18] The Deacons were Mr. Robt. Lachlison[19] and C. M. Quarterman.[20]

1879

On January 1st I received orders from Messrs. Duncan & Barnwell to turn over all property of theirs on Generals Island to Capt. A. S. Barnwell[21] as agent for Messrs. Duncan and Johnston of Savannah, Ga. Capt. Barnwell took charge immediately and employed me as his Overseer for this year, my pay to be $65 per month.

Jan. 5, 1879. We had a light snow here today; a lot of 7-day wonders, as I am told that snow has not fallen here in 20 odd years. The Rev. Albert B. Curry[22] became Pastor of the church here at the first of this year. There were the same officers, except that Mr. Pease died last summer.

In August Margie had bilious fever and convulsions. The convulsions were so severe that they caused paralysis of her right side from which she has never recovered. She is still left handed. She was certainly a sick child. While she was so ill, our physician's daughter was ill with diphtheria and died a few days later.

During this summer the proprietors built a house for me on Generals Island and on November 3rd we moved into our new home on the Island where we stayed all winter. This is the first rice crop I ever managed under Capt. Barnwell's supervision. We planted 217 acres and made 9,826 bushels, or 46 bushels per acre planted, which is an excellent yield.

The following is I think a correct list of the officers of the community at this date, both religious and civil: Bro. Ben Keyes[23] was Pastor of the Methodist Church, the Rev. Mr. Pinkerton[24] was Rector of St. Andrews Episcopal Church, the officers of the Presbyterian Church I have already given. City of Darien: Jas. Walker,[25] Mayor; Marshalls, C. H. Hopkins[26] and Alonzo Guyton.[27] McIntosh County: County Commissioners and Aldermen of the City of Darien: James Walker, Joseph Hilton,[28] Adam Strain, Isaac M. Aiken,[29] Jas. Lachlison,[30] Wm. H. Atwood,[31] T. H. Gignilliat,[32] Judge: Henry B. Thompkins.[33] Sheriff: T. B. Blount.[34] Clerk of Court: L. B. Davis.

Merchants—General Merchandise: [Editor's Note: Those in the following listing not now identified will be noted at later places in the Journal]. Adam Strain, Robert Strain, Wilcox & Churchill, J. & J. A. Walker, Hawes & Tyler, Jas. Larkin, Jos. Mansfield, Jas. Dancy, Charles Rothschild, Collat & Jacobson, Henry Miller, Sam Stern, Wanabaker & Weil, Philip Keller, John M. Fisher, John E. Heins, Jas. Cullinan. Druggist: W. H. Cotton & Co. Toys & Candy: Mrs. Todd. Lumber: Hiltons & Foster, Jas. K. Clarke & Co., D. M. Munro, L. Hilton Green, August Schmidt, R. K. Walker. Jas. E. Holmes, Young & Langdon. Practicing Physicians: Dr. R. B. Harris, Dr. Spalding Kenan, Dr.

Holmes.[35] Postmaster: D. Webster Davis. Lawyers: Wm. Robert Gignilliat,[36] L. E. B. DeLorme,[37] Walter A. Way, Henry Dunwoody. Timber Inspectors: Col. Barclay,[38] Ins. Genl., Wm. S. Mallard, H. S. Ravenel, Dean B. Wing, D. S. Sinclair, Octavius Hopkins, T. S. Quarterman, Chas. S. Wylly, A. C. Wylly, Jas. Winn.

I have gone into these matters this fully because it is interesting to some people to read such records and sometimes they are very helpful to those in search of such information.

McIntosh County salt marshes near Darien (P.S. Clark post card view, ca. 1900).

1880

January 1. I have been engaged again to overlook Generals Island under Capt. Barnwell's supervision.

Oct. 8, 1880. We had a heavy N.E. gale today with high tides and heavy rains. Generals Islands banks were broken in three places and a large amount of rice was wet. I was sick in bed with fever and Capt. Barnwell had to repair the breaks in heavy rains.

Dec. 30. Cold. Temperature 14º at 7 a.m., the coldest day since my arrival in Georgia. Owing the fact that it moderated immediately the orange trees were not badly hurt. But the cold broke the pump and the water pipes to the Generals Island engine[39] so that I had to take the pump to Doboy[40] to be repaired. This year we planted 307 acres and made 13,135 bushels of rice, or 42 bushels per acre of inferior quality owing to the heavy rains and the breaks in our banks, in the heaviest season.

1881

Jan 3. I was elected a Trustee of the First Presbyterian Church today, the other trustees being James Walker, D. S. Sinclair,[41] T. S. Quarterman and Robt. Mitchel. I was elected secretary of the Board, which office I held for 14 years. Then resigned it because the ladies of the Serving Society did not act fairly toward me and the congregation sustained their action. I have found churches, like other institutions, somewhat ungrateful for faithful service *freely* rendered.

With this year commences a new lease of Generals Island, which is to run five years, and I have been employed at $75 per month and $75 per annum for house rent in the summer season.

May 5. Fire! Our Generals Island mill, stables, straw shed and office were burnt. Fire was discovered bursting through the roof of the mill about 1 o'clock a.m. The cause of the fire will never be known but we suppose some hot ashes had been taken up into the second story of the mill the day before. This mill cost about $6,500 and was a total loss.

June 1. We removed to the Ridge[42] for the summer and occupied the house now owned by Mrs. James Lachlison on the point. Mother and sister Edith had arrived a few days before this, to stay a while with us. They stayed about three weeks and returned to Adams Run, S.C.

July 4. We commenced work on a new mill for Generals Island today. We bought a lot of machinery from a Mrs. Ashley near Jonesville[43] and carted it through the county to Darien. We bought a lot of lumber from Hiltons & Foster at Suttons Landing[44] near where we lived on the Ridge and framed the mill building there then rafted it to Generals Island.

This summer was so dry that the fire burnt in the straw at the Generals Island mill from May 5th until the middle of June before sufficient rain fell to out it...The rice was unusually small when the harvest water was put on and grub worms were troublesome.

August 27. Gale—We had a terrific gale from S.W. commenced about midday and got gradually worse until about 2 o'clock a.m. Sunday morning when it moderated somewhat. The wind blew about 60 miles an hour at its height. This is said to have been the worst gale since that of 1854. The damage in these parts was not great. Several stores in Darien were unroofed and the roads were rendered impassable. But it played havoc on Savannah river where the wind blew from N.E. driving the water in nine feet deep on the rice plantations. Over 300 negroes were drowned and a huge amount of property was destroyed.

Nov. 3. We moved back to Generals Island for the winter today. This year we planted 325 acres and made 12,625 bushels, or 38 3/4 bushels per acre planted.

1882

March 10. I bought today from Mr. Wm. Henry Ingram Lots 9 and 10 at Ashantilly[45] for $100 and went to work at once getting the lots cleaned up and ready for building on them. I also went to work putting up the part of the house that is now the L to the dwelling we now occupy and we built it in a hurry.[46]

Sept. 5. Fire in town! The stores occupied by Philip H. Keller, Robt. Levison, Mrs. Stewart and two stores of Collat Bros. and Adam Strain's warehouse were entirely destroyed this evening. It was a large fire.

Nov. 12. At a meeting of the congregation of the First Presbyterian Church of Darien held for the election of officers I had the highest honor obtainable to a layman put on me, that of being elected a Ruling Elder. This is the only office I ever really wanted.[47] Mr. A. Duff Curry was also elected to the Eldership and Messrs. D. S. Sinclair and A. E. Dimmock were elected Deacons.

Darien Presbyterian Church, Bayard Square, built 1876, burned 1899.

Dec. 31. The officers elected by the church on Nov. 12th, including myself, were ordained and installed into their respective offices and I took my seal in Session as a member thereof and, I think I may say, this was the proudest day of my life.

During this year we planted 257 acres and made 9,132 bushels of rice, or 36 bushels per acre.

1883

April 16. We moved to our summer place at Ashantilly, where we now live and we have not moved away from it since. I received orders from Capt. James H. Johnston of Savannah to take charge of Generals Island and everything on it and hold it in his name as Capt. Barnwell had been relieved of command and that henceforth I should report directly to and receive my orders directly from him only.

Oct. 12. I attended Savannah Presbytery at its meeting at Hazlehurst, Ga. today, as representative from the Darien church. This Presbytery dissolved relations existing between the Rev. A. B. Curry and the Darien church. Mr. Curry will go to Gainesville, Fla. and we are without a Pastor. Mr. Curry is a fine man and a fine preacher.

Dec. 31. During this year we planted 270 acres and made 8,375 bushels, or 31 bushels per acre, the poorest crop I had made up to this time.

1884

Jan. 8. Cold! Temperature 18º at 7 a.m. All oranges on the trees are ruined and some trees are killed. Other fruit trees are badly damaged. Nearly all of my fine peach trees were killed.

Aug. 10. Dick Wylly was killed by Lazarus Harris, both colored, on the Shell road[48] just opposite the negro cemetery[49] last night. I had to serve as coroners juror in the case, my only service in this capacity to the date of this writing (1900).

Aug. 31. Fire in town—Charles Rothschild's[50] store and Fulton's[51] butcher shop were burned last night.

Sept. 6. Capt. Bourke Spalding[52] accidentally shot and killed himself yesterday while hunting by himself. He was buried this evening in St. Andrews Cemetery by lamp light.[53]

Dec. 31. We planted 200 acres of rice on Generals Island this year and made 8,200 bushels, or 41 bushels per acre.

1885

March 4. Our 5th child, Emma Strain, was born at 5.15 o'clock p.m.

Aug. 16. The congregation of our church met today (I moderated the meeting) and elected the Rev. James N. Bradshaw Pastor. Mr. Bradshaw has been supplying our church for the last six months.

Aug. 24. Storm—We had quite a severe storm today. Barometer 29.40. This storm did not do us great damage but it wrecked Charleston, S.C.

Sept. 5. We made our first shipment of rice today. This is the first shipment of rice received in Savannah this season. It is the heaviest rice I ever saw, weighing 50 lbs. per bushel. It came from #4 and was white rice. The season so far has been very wet.

Oct. 11. Storm! After a series of very high tides and heavy rains, the weather culminated in a terrible storm from north east. Nothing like it since 1854. The water rose to an average height of 4 1/2 feet on Generals Island. All of our rice, except that stacked in the barnyard floated off and rafted in different parts of the place and some of the rice went on the banks and out to sea. It took two weeks hard work to get the rice spread out in the stubble so that it would dry. The whole of the Altamaha delta was a sea of water from Darien to Glynn County high lands. The loss was enormous. Butlers Island lost its total yield of 600 acres, except 300 bushels saved in a badly damaged condition. Generals Island lost fully 25% of the crop besides the great loss to quality. The water did not leave the place until the 13th.

Oct. 15. Mr. Richard L. Morris[54] died today at Inwood[55] of haemorrhage of the lungs at 8 o'clock a.m. Buried next day in St. Andrews Cemetery.

Dec. 9. An election was held today in McIntosh County to determine whether or not whiskey shall be sold. Result was Yes, 560! No, 47!!!

Dec. 25. We had a very pleasant basket picnic and oyster roast at the Ashantilly school house. It was a Presbyterian affair and was well attended and much enjoyed.

Dec. 28. This year we planted 235 acres and shipped 6,846 bushels, or 29 bushels per acre, notwithstanding our great loss by the storm of Oct. 11th.

The rainfall of this year was 63 in., the greatest on record.

During this year the artesian well on Broad Street in Darien was bored. It was originally 487 feet deep. It has since been deepened about 30 feet.

1886

Jan. 6. With Mr. Johnston's permission I enter a co-partnership with James Walker to plant Rhetts Island[56] for five years. I have a 1/3 interest and am to manage the outdoor affairs while Mr. Walker manages the finances.

March 23. Inverness Lodge #29 Knights of Pythias was organized in Darien tonight with 26 charter members. I joined and was their charter Prelate.

May 3. We hear today that Henry Todd[57] the rich colored man and highly esteemed citizen of Darien died in Atlanta, Ga. on the 1st inst. His remains were brought to Darien and his funeral was held in our church and he was then placed in his vault in the public cemetery.[58] He left the greater part of his property to the schools and churches of this community. The Presbyterian Church, of which his wife was a member and which he attended, received 10% of his estate, being $2,141.00. He was an unassuming man. He left a widow but no children.

July 15. There is a large freshet in the river, a rare thing. I am told that the last July freshet occurred in 1786, just 100 years ago. This freshet did not hurt Generals Island, but it entirely destroyed all upriver crops of rice, even Butlers Island losing nearly the whole crop. Gignilliat, Mallard,[59] Sinclair, Nightingale[60] and others lost about all their crops.

Aug. 21. Robt. Deurer (colored) was killed by lightning at My Hall Mill[61] today.

Aug. 31. **Earth-quake.**—At 9.25 o'clock p.m. we had the severest earthquake within the memory of man in these parts. It lasted about one minute. The people in these parts were more frightened than hurt. Some thought the final judgment had arrived, while others thought that ghosts had visited them. Those undressed "stood not in the order of their going" but left their beds or houses as they were. It will be long remembered by those who felt it. The sensation experienced is peculiar and by no means pleasant.[62]

No great damage was done here but Charleston, S.C. was wrecked by it and many persons were killed or wounded there. The Charleston & Savannah railroad was badly torn up and twisted. We got no mail for seven days, the mail agents being cut off at Charleston. We heard all sorts of reports but could not get anything definite until Sept. 4th when, having telegraphed friends in Savannah, they sent me newspapers by the steamer.

At that time Darien had no railroad and traveling had to be done in a round about way. Two steamers plyed between Darien and Savannah making two trips a week. These steamers were the *David Clark* and the *Centennial*. The *Hessie* was then also running between Darien and Brunswick.[63] Those who did not care to take either of these routes took a little steamer—a rowboat—to Hammersmith Landing[64] and took a hack from there to Sterling on the railroad. This last in the route we received our mail by. This was the quickest route but the steam boat was much more comfortable, for we left

Darien in the afternoon of one day and got to Savannah the next day, the steamer feeding and furnishing berths to its passengers.

But to return to the subject in hand—the Earthquake—from 9.25 p.m. of the 31st August until 11 o'clock I counted six other shocks and my watchman on Generals Island reported two others, the last at 5 o'clock a.m. Wednesday, which was accompanied by an electric display of considerable brilliancy. I was asleep when the first two shocks commenced, but wife waked me up before it was over and we left the house and outed all lamps. Before the shock it was still and hot, the thermometer being about 90º and the barometer 29.65. At 9.30 o'clock, five minutes after the first shock, the thermometer registered 78º and the barometer 29.85. We stayed up until 11 o'clock and then went to bed and, as we slept the rest of the night, we did not feel those shocks that occurred after 11 o'clock.

Steamboat Hessie made daily runs between Brunswick and Darien until 1918.

Sept. 1. We felt another shock, being the 9th, at 4.52 o'clock p.m.

Sept. 4. I made my first shipment of rice today. The 10th Earthquake shock occurred at 9.35 last night and it made things lively while it lasted.

Sept. 23. A spark from the smoke stack to the Engine started a fire in the Generals Island barnyard in the straw there and came near burning all the buildings down. We had a hard fight to save them but we did eventually confine the fire to the straw stacks which were all lost. The negro women fought the fire like Trojans. Our loss was small.

Sept. 25. J. T. Dent's[65] rice mill was burnt today about 11 o'clock a.m. This mill was on his plantation in Glynn County but in sight of Darien.

Oct. 7. Mrs. Lewis M. Bealer[66] died on the Ridge of typhoid fever contracted in Decatur, Ga. I am told that her death was largely due to fright from the recent earthquakes.

Oct. 10. Baby Emma, who has had cholera for some time, became so much worse today that wife and I took her to Savannah to Dr. Harris, our old family physician.

Leaving Darien in the afternoon, we took the steamer *David Clark*, and got to Savannah next morning. Emma improved rapidly under his treatment and the change of air and we returned home with her much better on the 13th.

Dec. 31.   During this year we planted 271 acres and made 10,300 bushels, or 40 bushels per acre.

1887

Jan. 1.   I have bought all of Mr. Johnston's property on Generals Island and have leased the place for three years at $800.00 per year.

I paid Mr. Johnston $666 for the mill and $834 for all other property. I have him three notes, one for $834 payable Jan. 1st, 1888, one for $833 payable Jan. 1st, 1889, one for $833 payable Jan. 1st, 1890. I have since paid the first two of these notes and the accrued interest at 7%, but the last has had only $200.00 paid on it, which payment was made in 1896.

Jan. 18.   Fire in town! The most disastrous since the war.[67] At 4.30 o'clock p.m. the Masonic Hall[68] was discovered to be on fire from a defective flue. That building was soon consumed. Then the Magnolia Hotel[69] caught and burned. Then the *Gazette* office[70] and the stores of S. A. Way and J. E. Heines, and J. E. Holmes'[71] office and three old unoccupied tabby buildings.[72] These were all burned to the ground. The wind blew hard from N.W. and so the rest of the town was saved. The Fire Engine did nothing worth mentioning. This fire made a huge opening in the town. None of the buildings have ever been rebuilt.

April 3.   Mrs. E. M. Blount, the largest woman I ever saw, died at 1.30 a.m. and was buried this evening in St. Andrews Cemetery. I acted as pall bearer. It took 12 of us to handle the coffin. She was so huge that no ready made casket could be found large enough and one had to be made. Then *this* casket would not go in the hearse so had to be taken in a spring wagon. She weighed either 285 or 385 lbs., I have forgotten which. She was a fine woman, much beloved.

April 30.   John Grant, a highly thought of and much respected colored member of our church, died today and was buried by the white people as a mark of respect.

July 26.   Mrs. Mary Ann Todd, wife of the rich colored man [Henry Todd] whose death is recorded [here earlier] and a much respected member of our church, died in Atlanta today. Her body was brought to Darien for burial and her funeral was held in the Presbyterian Church and her remains were then placed in their vault alongside of her husband. The white people took entire control of the funeral.

Mr. & Mrs. Todd and old John Grant were considered by our white people as dark-skinned white folks and they enjoyed the love and respect of the community to a remarkable degree. Mr. & Mrs. Todd were free born and I believe were slave holders. They were of "Gingerbread color." John Grant was a mulatto and *poor*. I think he was a slave. So it will be seen that property was not the only consideration in these cases. It was due to the way in which they conducted themselves. It was a question rather of what these people were than what they had.

Aug. 4.   We are advised that the largest freshet ever known on the Altamaha river will soon be down on us. At Macon this freshet is 32 inches higher than the celebrated Harrison freshet; and is 18 inches higher than the freshet of April 1886.

Aug. 10. The great freshet has reached Savannah and the rice places there are all flowed, some of them six feet deep. This freshet was the highest ever known in the Savannah river.

Aug. 14. The great freshet is here. Generals Island is flowed from 1 1/2 to three feet deep all over it. The entire delta is under water. I will state here that the crest of a freshet in the Altamaha river travels 50 miles in 24 hours and it takes 15 days from the time it reaches Macon to reach Darien, Darien being 750 miles from Macon by river.[73]

This freshet did not injure me much but all the other places in the river were more or less hurt.[74] On Altama, Wrights, Carrs and Cambers islands the loss was total. The Cathead[75] planters, Gignilliat, Mallard and Sinclair, lost heavily. Butlers Island's loss was almost total.

Aug. 15. The full force of the freshet is here and tonight's tide ran over the Generals Island flood gates for the first time in my experience. The place is flowed to the heads of the rice. We have had no mail in four days, the railroad [at Sterling] being under water for miles.

Sept. 8. Today makes 22 days the wind has blown continuously from N.E. Unusual.

Sept. 17. I have a large break in the Rhetts Island banks which I repair tonight.

Sept. 19. I have still another large break on Rhetts Island. I repair it tonight. About 20 acres of rice got badly wet.

Oct. 19. I and my family leave today by steamer *St. Nicholas* for a visit to upper South Carolina. We arrived at Cross Hill on the 21st.[76]

Nov. 11. By consent of parties Mr. James Walker and myself dissolved our co-partnership from two years ago to plant Rhetts Island. Lewis M. Bealer is buying my interest. I planted Rhetts Island two years and Mr. Walker's Cathead place of 30 acres one year and made fairly good crops, but my share of the profits for two years was only $126. So I gladly let it go.

Dec. 3. This year I planted 244 acres—144 acres for money and 100 acres on shares with the negroes. I shipped 6,834 bushels and saved seed for 30 bushels an acre, which is a poor yield, but I got fair prices and made some money.

1888

Feby. 6. I bought a Whitehall boat today from Capt. Birdsall of the schooner *Varina*.[77]

May 1. I commenced work on the new front to our house at Ashantilly. I will take down the dwelling on Generals Island and put it up there [at Ashantilly], thus making the original building an L.

May 8. My flat sunk in Snows Creek[78] with a load of bricks and lumber for my new house. I got it up again with considerable trouble that afternoon. I lost nothing by the accident except the extra cost of labor raising the flat and unloading and re-loading the flat.

Sept. 26. A freshet has passed without doing any harm. It was nothing like as large as those to the north of us, but it caused considerable loss to the few upriver planters who planted this year. The summer freshets of the past two years have done them so much damage that Couper on the Corbin places[79] and Nightingale on Cambers Island[80] have stopped planting. The freshet ruined the Gignilliats so that one by one the rice planters are knocked out.

Nov. 25. My watchman, Tom Lawson (colored) is missing. His hat, keys & *whiskey* were found in my boat which was adrift this morning.

Dec. 7. Lawson's body was found today. He had—we suppose—fallen out of the boat in a fit of drunkenness and swam to the rafts,[81] from which he tried to get to land on Generals Island, but he bogged and must have laid down and gone to sleep and the tide rose over him and drowned him. When found he was opposite Darien on the Generals Island shore bogged to his waist.

Dec. 9. Mr. W. W. Churchill died at the residence of his daughter, Mrs. W. A. Wilcox of Ashantilly Nov. 28 and was placed in a metallic casket and kept until today in order that a vault, which was being built, might be finished before the interment of his body. His funeral was held today and I acted as a pallbearer. We placed his remains in the vault at St. Andrews Cemetery.

Dec. 31. I planted 275 acres of rice this year, 175 acres for money and 100 acres in shares with the negroes. I made 6,400 bushels, an average of 29 1/2 bushels per acre, which is poor.

1889

Jan. 3. I acted as a manager of an election today for offices for McIntosh County until 2 a.m. this morning and again from 8 until 11 a.m. I have no desire to repeat this experience.

Jan. 12. Tom Hopkins, about 12 years old, son of Mr. Octavius Hopkins[82] accidentally shot and killed himself in a boat at Mr. Strain's landing on the Ridge today. He and several other boys were going hunting and he, in moving the guns in the boat, was shot, the load passing through his body and killing him almost instantly.

July 30. I drove a pump at Ashantilly today and found water at 17 feet. I drove 20 feet but had to draw the pipe up to 17 feet as no water was found below 17 feet.

Aug. 27. Our sixth child, and third son, James Houstoun Johnston, was born at Ashantilly today.

Sept. 25. I went to Savannah yesterday. In returning today some of the Str. *David Clark*'s machinery broke and we laid up at Warsaw [Wassaw] Island several hours.[83] I visited the beach and the beach pavilion while waiting there.

Oct. 8. Steamer *David Clark* was burned at Fernandina this morning.

Oct. 18. The Rev. N. Keff Smith,[84] who has been called to the pastorate of our church, arrived with his family by Steamer *Hessie* today.

Oct. 20. I was elected Clerk of Session today and have held that honorable office ever since.[85]

Nov. 4. Mrs. John M. Fisher died at 3.45 o'clock a.m. She was buried that evening at St. Andrews Cemetery. She died at Ashantilly of a complication of diseases.

Dec. 3. I am trying a new machine, the Engelberg rice huller. I am so much pleased with it that I buy it [for] $250, which is a special introductory price.

Dec. 30. I am on the Grand Jury. This year I planted 286 acres of rice and made an average yield of 27 bushels per acre and lost money, which embarrassed me considerably, so much so that I did not succeed in making arrangements to plant until Feby. 1, 1890.

1890

Jan. 13. Mr. James E. Holmes, Inspector General of Timber & Lumber [for Darien], and a particular friend of mine, died today and was buried next day in St. Andrews Cemetery. A good man who drank himself to death!

Jan. 31. After much trouble, I succeeded in making arrangements by which I will plant Generals Island another year. I have leased the place for three years—the first two years at $700 per year and the third at $750. Messrs. Louis Collat, J. H. Johnston, Joseph Mansfield[86] and Henry A. Weil[87] have endorsed for me thus enabling me to get money to continue my planting operations.

April 30. I had Halvorsen the photographer take a photograph of my family as a group. We could not keep baby Houstoun quiet so had to have his photograph taken separately.

May 8. The McIntosh County Sunday School Association had their annual picnic today at what is now called Eulonia,[88] where the stage road crossed the Darien Short Line Ry.[89] We went and found almost everybody there. In the afternoon all who cared to go were invited by Pres. R. K. Walker of the Ry. to take a ride over the road. Mr. Walker had a train of about six flat cars with cross ties placed crosswise for seats ready and we were taken out to the northern terminus about six miles away, then to the southern terminus at Crescent[90] and Belleville[91] then back to the starting point. The trip was greatly enjoyed.

May 30. Str. *St. Nicholas* was burned in Savannah today.

May 31. A white man, Fred Graves, was found dead in his house at the Thicket[92] this morning.

July 2. W. C. Wylly's[93] rice mill on Broughton Island[94] was burned yesterday afternoon.

July 9. We had a family excursion to Egg Island[95] on the little steamer *Carrie A. Ward* today. Among those who went were the families of Rev. N. Keff Smith, H. A. & Simon Weil and mine.

Sept. 18. I bought a black mare from Mr. John Stebbins[96] of Riceboro today for $135 and named her "Jessie."

Sept. 29. The heaviest rain I ever saw has fallen in the last 24 hours, something over 10 inches of water, flowing the rice fields.

Dec. 10. J. K. Clarke's[97] fine office and dwelling [in Darien] were burned this morning.

Dec. 31. During this year we had a rainy spell of six weeks in August and September that ruined nearly all the rice in Georgia. I saved 37 bushels per acre (I cannot recall the number of acres planted) and sold my crop for from $1.37 to $1.42 per bushel and cleared $3,700.

This being the end of another decade I will try to show from memory who were in the different offices, merchants, &c. in the town and community.

Church officers: Presbyterian Pastor–Rev. N. Keff Smith. Elders: R. K. Walker, T.

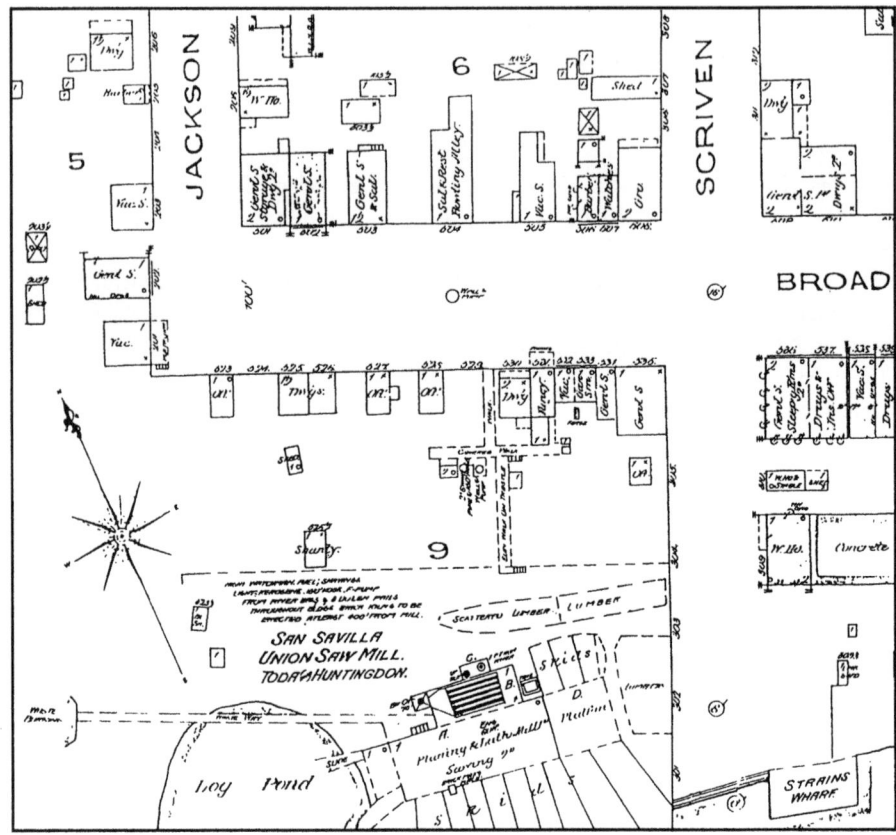

Broad and Screven street intersection, Darien, from Sanborn Insurance Map, 1885.

S. Quarterman, James Walker, A. Duff Curry and myself. Deacons: C. M. Quarterman, J. A. Walker,[98] D. S. Sinclair.

    Pastors—Episcopal: Rev. John W. Motte[99] and Methodist: Rev. Thomas E. Davenport.

    City of Darien—Mayor: James Walker. County Commissioners and ex officio Aldermen of the City of Darien: James Walker, James Lachlison, Allen McDonald, Dr. Spalding Kenan,[100] J. M. Fisher,[101] W. M. Young, H. A. Weil.

    Post Master: Thomas A. Bailey. Collector: R. W. Grubb.[102] Judge: Robt. Falligant. Solicitor: W. W. Fraser. Sheriff: T. B. Blount. Clerk: S. A. Way. Ordinary: Wm. J. Donnelly.

    Doctor: Spalding Kenan. Lawyers: C. L. Livingston,[103] W. A. Way and Wyatt de R. Barclay.[104]

    Hotel & Boarding House: Mrs. Julia F. Palmer, Mrs. Susie Way.

    R. W. Grubb has been proprietor and publisher of the *Darien Timber Gazette* ever since my arrival in these parts.

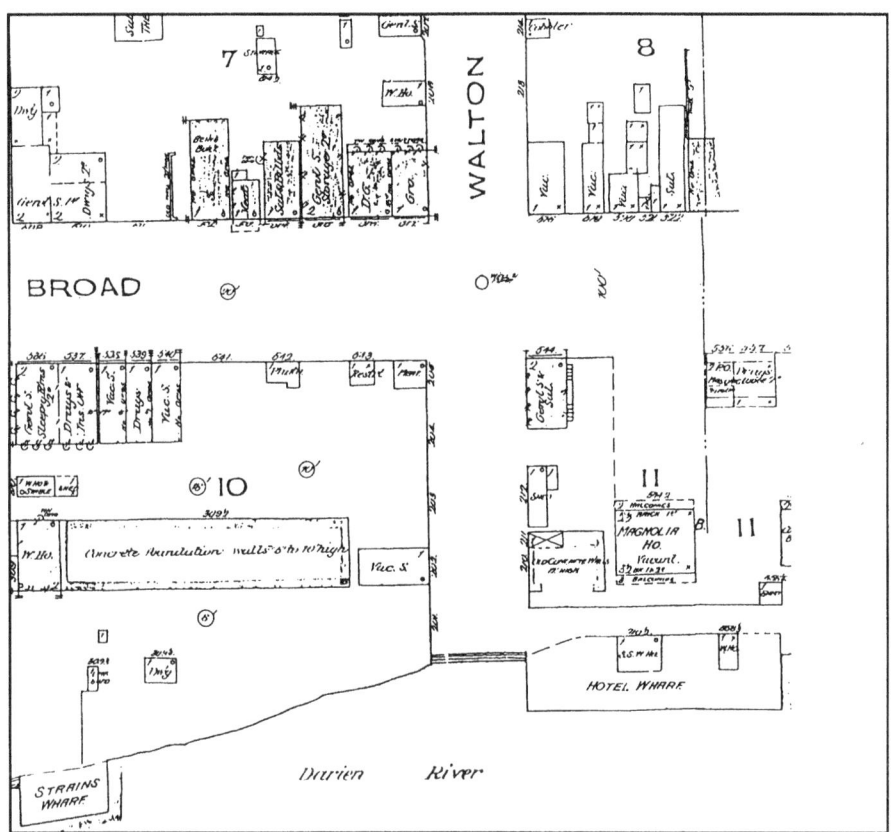

Broad and Walton street intersection, Darien, from Sanborn Insurance map, 1885.

1891

April 13. I bought a double-seated buggy and one set of harness today from J.A. Walker for $93.00.

April 27. I go to Savannah to see the great spectacle, Paine's "Destruction of Pompeii." I take Claudie with me. She will spend several days with Capt. M. P. Moina's family by invitation. I returned home by way of Hammersmith Landing where my rowboat was to meet me. Mr. William Konetzko[105] was with me. When we got there, through some mistake my boat was not there so we were in rather a bad fix. We found an old bateau[106] nearly in two lengthwise, an old jug bottom and a clapboard and so we put the bateau down and I paddled while Mr. Konetzko bailed the boat with the broken jug bottom and in this way we crossed to Champneys Island[107] where we got Capt. A. S. Barnwell to send us to Darien. We got to Darien about 11 o'clock at night. The trip laid Mr. Konetzko up for several days.

May 21. I went to Brunswick to attend the meeting of the Grand Lodge Knights of Pythias as representative from Inverness Lodge #29 at Darien. I stayed at the Oglethorpe Hotel,[108] which I found the most comfortable homelike hotel I ever stopped at. While in Brunswick I visited about every part of the town and saw all that was to be seen.

July 3. There was an excursion given today from Darien to St. Simons Island under the auspices of the McIntosh Light Dragoons. I and my family went. We visited the Hotel and bathed in the surf and had a good time generally. My last surf bath before this was on Long Island, now called Isle of Palms, near Charleston, S.C. in 1873.

Sept. 7. The negro reported with smallpox July 10th has since died. Today three other cases were discovered on the outskirts of Darien, making four cases to date. The community is much worked up over it. I and my entire family was vaccinated.

Dec. 31. During this year I planted 243 acres on Generals Island and made 7,103 bushels, or 29 1/4 bushels per acre. Rice I cleaned here netted me $1.37 per bushel. That sent to Savannah netted me 98 1/2 cents, so that I lost money on the crop.

1892

Jan. 10. The colored members of the Presbyterian Church of Darien were organized into a separate church called "Grant Chapel." J. D. Taylor, a colored Licentiate, preached for them.[109]

Feby. 7. Fire in town—The old Long warehouse was set on fire by some unknown person near midnight last night and burned down. Then C. M. Quarterman's warehouse was burned, the Fire Engine doing no good.

Feby 13. Fire in town—Some unknown person set fire to the offices of Todds Mill,[110] burning that, then J. A. Walker's store on the S.W. corner of Broad & Walton streets,[111] then the Todd residence, then several small offices on the south side of Broad St., making a huge opening in that part of Darien. The fire was discovered about 10 o'clock p.m.

As usual, the Fire Engine did very little useful work. The Engine is a fine one but, through the carelessness of the authorities, it is always out of working condition when wanted. Late at night the engine got to work and probably saved the western portion of Darien from destruction.

March 4. Fire in town—The McIntosh Academy was set on fire by an incendiary—still unknown—about 9 o'clock p.m. and burnt down. The fire engine was disabled through carelessness. A barrel of water would have outed the fire when it was discovered, but even that little could not be had. This fire has caused a heavy loss on the community. The building had recently been thoroughly overhauled and enlarged and for the first year since the war we had a really good school. Now all is undone. For the present the Courthouse will be used for school purposes, until the Academy can be rebuilt.

March 7. Several attempts have lately been made to burn houses in different localities in Darien and the whole community is stirred up because of them. Large rewards have been offered, but so far no one has been arrested. The rewards offered aggregate $800.00. At Ashantilly we patrol the roads every night, all night.

March 26. Rhetts Island mill was accidentally destroyed by fire this evening about 4 o'clock. The loss was quite heavy, though covered by insurance.

June 28. Mrs. John Muller[112] died at her home on Harris Neck[113] today.

Aug. 22. I commence harvesting today on Generals Island. Prof. J. W. Flynn, D.D. of South Carolina College is visiting me. He is fishing at the Wolf Island Club House[114] but by invitation, he came up and preached for us yesterday. I sent him back to Wolf Island today. He is a fine man and a fine preacher.

Oct. 9. The Rev. N. Keff Smith preached his farewell sermon today. He goes to Ebenezer Church, Charleston, S.C. He leaves because we are too poor to pay him a salary sufficient to support his large family. The parting is with great regret on both sides. He is a fine man and a great singer—in fact a composer of music—he is full of fun and a general favorite in the entire community.

Oct. 21 [sic]. Is being celebrated over the entire U.S. as the 400th anniversary of the discovery of America by Christopher Columbus.

Nov. 19. Fire in town—The stores of J. D. Wilcox and A. H. Collat on the north side of Broad Street, western end of the street, were burned about 10 o'clock last night. The fire was started in Wilcox's store and was supposed to have been of incendiary origin. The Fire Engine did some good service here and probably saved several nearby buildings.

Nov. 25. Broughton Island mill was burned this morning for the second time in three years from what is supposed to be accident or carelessness.

1893

Jan. 14. Still colder—Temp. 17º—the coldest since 1886. The orange trees are not badly hurt, which is strange.

Feby. 15. Philip Maxwell, a much esteemed colored man, and the first and only Ruling Elder of Grant Chapel, the colored Presbyterian Church here, died yesterday and was buried today.

March 25. The Rev. Henry K. Rees[115] the Episcopal Rector, died in Darien this morning after a protracted illness. He will be buried in St. Andrews Cemetery today. His father was Col. Ebenezer S. Rees for 22 years an Elder of our church long before the war.

April 1. I have been engaged by Mrs. F. B. Leigh[116] to take charge of Butlers Island. The negroes will plant the place, renting what land they wish. I am to keep the place

Timber ship loading on Darien River, ca. 1900.

in repair and collect her rent and have general oversight of everything on the place. She will pay me $20 per month.

May 9. The McIntosh County Sunday School held their annual picnic today at Moore Lake, about 12 miles from Darien near where the Seaboard Air Line [railroad] crosses the Altamaha River. We saw sand hills out there from 40 to 50 feet high and gopher holes ever so deep. We had a pleasant day but caught no fish. My whole family, and about everybody else in the county, went to it.

May 25. Edward Poppell was shot through the head from behind by a young white man named Branham in S. A. Weil's barroom. He recovered but lost the sight of one eye. It was a cowardly thing but Branham was never punished for it—I do not know why.

July 1. I am appointed by Judge Falligant—the presiding judge of the circuit—a member of the board of Academy Commissioners to fill out the unexpired term of Mr. Jas. Walker, resigned. I received my commission as such today and attended a meeting of the board which is at present comprised of Dr. Spalding Kenan, Chairman, H. S. Ravenel, Secretary & Treasurer, Capt. A. C. Wylly, Mr. Octavius Hopkins, Capt. Wm. Henry Atwood, Mr. Jas. Lachlison & myself.

July 4. First heads of rice were seen today. The weather is very hot. I have cleaned on my huller since Sept. 1st, 1892 the following amounts of rough rice: For myself 1452 bushels. For others as toll 1758 bushels. I have received as toll for that cleaned for others $246.85 of which about $165.00 is profit.

July 29. Times are very hard—in fact, we are in the midst of a season of great business depression all over the civilized world. No one seems to know just what is the matter but money is not to be had. Banks are failing in different parts of the country every day...Hog products are high, but almost everything else is low—cheaper in fact than was ever known before in this country...Mrs. Leigh writes me from England that a drought is severe there, that hay sells for $50 per ton. My crop is poor and my prospects are poorer for prices are far below the profit mark in rice.

I will make an effort to quit rice planting after this year. It is hard to have to give up a pleasant occupation followed for 15 years, but I see no help for it.

Aug. 12. Yellow fever has broken out in Pensacola, Fla. and also in Brunswick, Ga. Darien has quarantined strictly against both places, as is usual with us in such emergencies.[117]

Aug. 14. The news from Brunswick is that U.S. Surgeon Branham and one other person there has the yellow fever and that the people are leaving there in large numbers, panic stricken. The doors of the railway coaches have to be locked when the coaches are filled in order to keep them from being overcrowded, and even then, the people crowd in through the windows. Seven coaches filled to overflowing left there last Saturday and 11 coaches left yesterday, and still they go!

The steamer *Hessie*, which makes daily trips between Brunswick and Darien, has been taken off the line because she is not allowed to land at Darien.

Times are harder than ever...The Darien Bank[118] claims to have money to lend, but they require about $5 worth of security for every dollar loaned, and no one understands just why the panic exists.

Aug. 26. So far there have been two deaths from yellow fever in Brunswick, one of these being U.S. Surgeon Branham. The better class of people, at least those in comfortable circumstances, have all left the place, leaving the poor whites and the negroes only, in the town. These are suffering for the necessities of life and the city authorities are asking outside aid for them.

All of the coast cities and towns have put on rigid quarantine against Brunswick and the city is surrounded by a cordon of U.S. quarantine guards who will allow no one to enter or leave the place without permission of the U.S. Surgeon there in charge. It is difficult to travel, all travelers being required to show clean certificates of health, or be placed in quarantine. These things all add to the hard times and the money stringency continues. We depend entirely now on Savannah steamers for freight and passenger service.

Aug. 28. Cyclone–A terrible cyclone passed by yesterday...A heavy rain fell almost continuously all yesterday and last night...Although the storm blew from N.W. the tides were immense, covering Generals Island four to five feet deep and rafting 106 acres of rice that I had just finished harvesting, that is, cutting, tying and stacking in the fields. A large amount of my rice was carried out over the banks and to sea by the receding tide. I estimate that I lost $1500.00 worth of rice by this storm. I was the only planter that had cut any rice on this river and so I was the only one who lost any rice. In this sense, "the early bird did not catch the worm," but was caught itself instead...

The storm blew down two negro shanties on the Shell road and one of my shanties on Generals Island was blown off the blocks and my engine smoke stack blown down. There are no casualties in our immediate vicinity...Rhetts Island has one large and several small breaks...Butlers Island has a large break and several buildings badly damaged. Broughton Island has several large breaks. The rain fall was 5.70 inches. J[oseph] Mansfield's barn at the Thicket was wrecked, breaking several valuable farm implements.

The damage done on the South Carolina coast is dreadful, the loss of life at Beaufort and in its vicinity being about 1000 persons and the property loss is immense.

Savannah estimates her loss at $100,000. Charleston puts hers at $110,000. Savannah reports 20 lives lost. The telegraph wires are all down and news from the outside world is meager.[119]

Sept. 2. The telegraph wires are not yet repaired but the losses are becoming apparent from the storm just mentioned. The rice crop to the north of us is damaged about 75%. My loss is the heaviest on this river because I had cut more rice than any other planter here. A great many vessels were wrecked, among them the steam ship *City of Savannah*, from Boston to Savannah, which was wrecked on the coast of South Carolina near Beaufort entrance; but no lives were lost there. Two vessels from Darien were lost but I think the crews were saved.

Sept. 11. There having been so far only two cases of yellow fever in Brunswick and no new cases in the last 15 days, the Surgeon General ordered the quarantine raised last Saturday. The steamer *Hessie* made a trip from Brunswick to Darien today for the first time since the yellow fever scare commenced.

Sept. 14. Yellow fever has re-appeared in Brunswick and quarantines are in full force again. There are three new cases and two deaths yesterday.

Oct. 4.   Yellow fever is prevalent in Brunswick. There have been 70 cases and nine deaths so far.

Oct. 15.   D. S. Sinclair's turpentine distillery at Sidon plantation[120] was burned last night. The fire is thought to have been of incendiary origin.

Oct. 19.   I have now threshed all my rice on Generals Island, being 3,330 bushels from 204 acres—16 1/4 bushels pr. acre! Very poor!

Dec. 2.   I have today leased Rhetts Island from Mr. Jas. Walker and will plant it next year. I will plant 165 acres. The rental is $400.

Dec. 5.   The quarantine against Brunswick because of the yellow fever is raised today and the Str. *Hessie* has resumed her regular daily trips today.

Dec. 25.   The new railroad—the Florida Central and Peninsular Ry.—ran their first passenger train into Savannah today. Our nearest station is named Barrington.[121] Several Darienites took this train to Savannah at Barrington today. It will certainly fill a long felt want.

1894

Jan. 24.   James Walker & myself took a trip on the Darien Short Line Ry. to its terminus—seven miles this side of Walthourville to confer with the Rev. E. W. Way who has been called to the pastorate of our church.

Feby. 15.   I am learning to ride a Bicycle. I am getting along finely, but I seem to furnish a good deal of amusement to my family who usually look on while I am practicing.

Feby. 26.   Str. *Bellevue* broke her shaft on her return trip to Savannah last Saturday and we are now without communication with the outside world except by Str. *Hessie* to Brunswick.

West end of Broad Street, north side between Screven and Jackson streets.

May 4. The Presbyterian Sunday School held their annual picnic at Altama Plantation[122] today. I managed everything (and got a good deal of fault found with me for my pains). We took the steamer "Black Hawk" and two large row boats loaded with people. We landed at Hammersmith Landing and walked to Altama, a distance of only about one half mile. We left at 3 p.m. and took a trip down the south branch of the Altamaha to a point some distance below Broughton Island dwelling house and then turned back, going to Darien through Woods and Generals Cuts, arriving at Darien about 6 o'clock p.m. I think the day was generally enjoyed.

May 9. The old dwelling at Kells Grove occupied by Mr. James S. Townsend[123] was destroyed by fire today.

Last Monday a change was made in the mail route. Before that date the mail for Darien was put off at Stirling and was carried from there to Hammersmith Landing by buggy, then brought from Hammersmith to Darien by rowboat. Now our mail is carried by the Florida Central & Peninsular Ry. to Barrington (Now called "Cox") and is brought from there 11 miles to Darien by buggy twice daily. We now get the Savannah papers at 11 a.m. It is the best mail arrangement Darien has ever had, but there is room for improvement still.

May 17. The Western Union Telegraph Co. ran its line into Darien (via Barrington) today. Heretofore we have only had a private line to the outside world.

July 6. Licentiate T. M. Hunter[124] of Nashville Presbytery who has just graduated from the Southwest University in Clarksville, Tenn. and who has been called to the pastorate of our church has notified us that he will arrive tomorrow to take charge as pastor-elect.

Aug. 11. The hottest day of the season. Temperature 102° and but little breeze.

Aug. 14. The new steamer *Vigilant* built for the inland route between Savannah & Darien made her first trip today arriving here this evening.

Nov. 1. Our pastor-elect, the Rev. T. M. Hunter (who has been making his home at my house) and Miss Sallie K. Owen of Clarksville, Tenn. were married at that place at 11 a.m. today.

Nov. 5. Yesterday, I sent in my resignation of all the offices I held in the Darien Presbyterian Church because of adverse criticism and discourtesy on the part of the Ladies Sewing Society of the church. Our people—or some of them—have been "picking at me" ever since the picnic of the Sunday School held May 4th last, of which they appointed me sole manager and I think treated me so shabbily that I have concluded never again to have anything to do with the management of any frolic under the auspices of either the church or Sunday School if anyone other than myself has so much as one cent invested in it.

The offices I hold are Ruling Elder, Clerk of Session, Trustee and secretary & treasurer of the Board of Trustees. This last named office is the one I am especially anxious to get out of, having held it for 14 years continuously and in its administration made enemies and got no thanks for my trouble. Preliminary action was taken on my resignations yesterday and I was excused from service until the congregation can act. And they have called a meeting of our congregation to consider the matter next Sunday.

Nov. 7. At the general elections held yesterday, the Republican party has swept the country. Too much Democratic foolery. The term Democrat seems now to be synonymous with fool and the term Republican has been for some time synonymous with scoundrel.

At a meeting of the Trustees of the Presbyterian Church held yesterday my resignation as secretary and treasurer was accepted and J. A. Britt[125] was elected to that position. It is a thankless job and I am glad to be out of it.[126]

Nov. 28. The Darien & Western Ry. opened a General Managers office on Broad St. in Darien today. The sign seems out of place as the road is still 12 miles away.[127]

Dec. 29. Temperature 12° this morning and remained near freezing point all day although the sun shone. This is the coldest spell on record in these parts and will undoubtedly do considerable damage, especially to the orange trees.

1895

Jan. 1. I commence the year with a disaster. My best mule "Lou" got in a ditch on Generals Island last night and froze to death before she was found this morning.

This year I will plant Generals Island again. Rhetts Island proved an expensive venture to me. My losses there on the crop of 1894 will foot up something over $2,700...This is my worst disaster yet in my planting experience.

From this time on, until Oct. 2nd, 1898 my memoranda must of necessity be brief and incomplete because my book concerning that interval were all [lost] on Champneys Island in the great tidal wave of that date that destroyed my office and all other buildings on the place. Will get such data as I can from other available sources, such as my letter copies, the *Darien Gazette*, &c. &c.

January 26. The Darien & Western Railway was finished to Darien today and a construction train ran to the site of the present terminus. There is a great rejoicing in the community, most people thinking that Darien will immediately become a "boom" town. It would be a difficult matter to buy real estate here just now at reasonable prices. A photograph was taken of the Engine which was literally covered with people of the community.

Feby. 15. We had about two inches of snow last night which is the heaviest fall ever seen in these parts. The snow lay on the ground until about 2 p.m..

March 11. The first passenger train on the Darien & Western Ry. left Darien at 8 a.m. today for Darien Junction.[128] I bought the first ticket they ever sold and I have it now (1902).

April 15. The great negro preacher "Doctor" Robert Miflin of Darien—Baptist—died suddenly out in the country about two miles today and there is a great lamentation around the negroes. His body was embalmed and kept until April 28th when he was buried, the darkies arriving from Brunswick, Savannah and even more distant points. They gave him a big blow-out.

Dec. 19. The cypress mill on St. Simons Island was entirely destroyed by fire today, the property of the Hilton & Dodge Lumber Co. of Darien.[129]

Dec. 31. During this year I managed Butlers Island for Mrs. Leigh and planted

Generals Island on my own account, but did not clear any money. I came out just about even, maybe a little behind.

1896

Jan. 17. I received a letter from T. M. Cunningham, cashier of the Central Railroad Bank of Savannah, Ga. asking if I was open to an offer of business...I have agreed to manage Champneys Island for the Central Railroad on the following terms, namely, they are to furnish the place, equipment and money to run it, free of any expenses to me and, on my part, I am to manage the island, planting about 2/3 of the place for which I am to receive one fourth of the profits arising from the business.

Jan. 18. I visited Champneys Island today and presented an order to Capt. A. S. Barnwell from Mr. T. M. Cunningham, ordering him to transfer the business to me. I took formal charge of the place today and commenced work on the place at once,[130] hiring Joe Brown as foreman at $30 pr. month.

Feby. 2. The Knights of Pythias gave a fine entertainment at night. The exercises commenced with music and speeches in our fine new Lodge room over Mansfield's store. At about 11 o'clock p.m. the Lodge and invited guests went over to the Armory[131] where a fine supper was served. It was one of the "swellest" events in the recent history of Darien. The managers, as near as I can now (1902) recollect were T. A. Bailey,[132] R. P. Paul, A. E. Dimmock,[133] Henry A. Weil, myself and I think two others.

Feby. 4. Gov. W. Y. Atkinson visited Darien today as the guest of Representative Joseph Mansfield, and Darien had the pleasure of seeing and shaking hands with the Governor of the great state of Georgia! One youngster about 10 years old who was very anxious to see a *live* governor, remarked that he "looked just like any other man."

Sept. 10. The Champneys Island mill was burnt by a spark from the engine at 4:30 o'clock this evening. I had 2,900 bushels of rice in the building of which we saved about 1,300 bushels in good condition and 500 bushels too poor to ship. The machinery was saved in fairly good condition so that our loss will be covered by insurance. The loss of the mill building is a considerable inconvenience just now. We worked all night putting the fire out and saving the rice.

September 29th, 1896.

Storm—We had a terrible storm today. It came up suddenly and unexpectedly. At 8 a.m. the wind blew about 15 miles an hour from S.E., the barometer being about 29.50 and a light rain was falling. The wind gradually rose until 9:30 a.m. when it became so strong that we could not walk against it and a man on one of the tug boats[134] lying at the wharves sent me word that the barometer had fallen to 29.30 and was *wavering!* It is very seldom that a barometer ever wavers on land and I would have liked to have seen it, but it was impossible to get to the boat.

From 10 to 11 a.m. the wind must have blown fully 90 miles an hour and a rain fell all the time. The rain water did not seem to fall—it was driven level with the earth. This storm was something awful. It seemed as though nothing could withstand it. It was certainly the hardest wind I ever was in and had it lasted—say three hours—I don't think a building in Darien would have stood it. I saw some funny things but I did not think them funny until afterward. I was in Wolfsons store, one of the A. H. Collat stores, on the north side of the street. It was funny to see Capt. J. M. Holmes[135] trotting on *one* foot ahead of the gale with a large roll of roofing rolling just behind him hurrying him along.

Mc Dunwody & Tom Bailey stood under the huge oak by Strains store during the entire gale. Mc tried to look around the tree to see what the gale was doing when the wind blew the rim entirely off his straw hat. There was a horse and buggy tied in front of the store I was in. The buggy danced a jig, sometimes standing on one wheel. A slate off Strains store whizzed by me as I stood in the door and came near hitting me in the face. If it had, I guess I would not be writing now. Soon after 11 a.m. the gale moderated and I started for home at Ashantilly. I got nearly to Wilcox' corner when a sudden gust struck me and I could not get but one foot on the ground and went dancing around on one foot for several seconds. There were 22 pine trees across the shell road. We found our families more scared than hurt, or I should say badly scared, but unhurt. By 1 o'clock p.m. the gale had entirely subsided.

Results of the gale.

Champneys Island–17 out of 25 buildings were blown down. Generals Island–All my sheds were blown down, badly damaging some of my implements and machinery and about 50 sacks of my rice was dumped in the river. I recovered most of it in a damaged condition. Butlers Island–Dwelling badly injured, the feed barn blown down and two negro houses. Crop badly damaged.

Darien–The streets are impassable because of fallen trees. St. Cyprians Church (Episcopal) and the negro Methodist church are blown down. The white Methodist church is badly twisted, the Presbyterian church is blown out of plumb, Mansfields store is wrecked, the Knights of Pythias Hall above Mansfields store is ruined, nearly every other building in town is injured more or less.

In the surrounding county the turpentine industry is ruined, so many of the pine trees are blown down. I do not remember that anyone was killed but there were some very narrow escapes. I do not think a single tin roof in the community remained after the gale. All were blown off. I had two schooners loaded with rice on their way—one to Charleston and one to Savannah. Both escaped injury.

Oct. 9. This year I planted Generals Island on my own account and Champneys Island on shares with the Central Railroad and managed Butlers Island for Mrs. Leigh. At the end of this year she gave me my discharge and put Capt. A. S. Barnwell in charge.

October 21st, 1896. Claudia married Jesse Alexander Britt tonight. The ceremony was performed by the Rev. T. M. Hunter in the Presbyterian Church in Darien. After the ceremony a large number of friends gathered at my home where we served a good supper which was apparently enjoyed by all.

1897

March 3. Charles H. Hopkins was coming into Darien last night on a crank car on the Darien & Western Ry. and ran into something and was thrown off, injuring his spinal column to such an extent that, although he lived about two years, he was unable to help himself until his death.

April 1. An unusually large freshet is down on us, said to be the largest since the great Harrison freshet of 1834. Champneys Island is entirely under water. A N.E. gale is also blowing which has flowed up the settlement two feet deep. It is the highest I ever saw. The difference between high and low water on the upper end of Champneys Island was only 4 1/2 inches. At the settlement the fall was about 2 1/2 feet.

About the first of April the new Ice Factory in Darien commenced work. It is quite an acquisition, and convenience also, to the entire community.

May 8. Mr. Adam Strain, probably our foremost citizen and my first friend made in Darien, died at his home on the Ridge early this morning of pneumonia. He will be buried at St. Andrews cemetery tomorrow morning. He is a man that will be greatly missed in our community. He was not a Christian but he was a liberal man. He did many charitable acts of which the community was not aware. He was a consistent man; and his yes meant *yes* and his no meant *no*.

Aug. 9. The Bell Telephone Co. completed their long distance telephone line to Darien today and it was opened for business so that we are now connected with Savannah, Brunswick, Jacksonville and even New York!

Oct. 1. Our pastor, the Rev. T. M. Hunter, resigned the pastoral charge of our church some time ago. He leaves today for his new field of labor in Trenton, Tenn. We are sorry to have Mr. Hunter leave us. (He later regretted the move, when it was too late).

1898

Jan. 18. My first grand child, Charlotte Legare Britt, was born at about 5 o'clock this evening.

February, 1898. The U.S. Battleship *Maine* was blown up in Havana (Cuba) harbor and the country has gone wild over it. It is thought that the treacherous Spaniards did it through revenge because of the deep sympathy of the American people for the Cubans who have been struggling for their freedom for several years. An investigation is to be held to place the responsibility for the dastardly act by which 260 odd men lost their lives.

June 19. About sunset this evening there was a terrible boiler explosion at the Lower Bluff mill. Mr. Charles McCosker[136] was killed before help could reach him and Charles Desverger and a negro boy were so badly burnt that they died within two days.

Darien, Ga., October 5th, 1898. We had a terrible storm on last Sunday—the 2nd inst. accompanied by a tidal wave of immense proportions.[137]

The consequences and effects of this storm were awful—some of which are as follows.

There were 32 persons drowned. On Champneys Island there were 14 persons, of whom seven were drowned. These were Rosa Holmes and her two little daughters, Mary and Martha, twins, and Carrie Magwood & baby, Doll Simmons and the night watchman James Brown. Only one of the bodies have been recovered so far, that of little Mary.

At Ashantilly the tide rose 8 1/2 feet in about two hours from 11 a.m. to 1 p.m.

Darien Ga. October 5th 1898

We had a terrible storm on last Sunday the 2nd inst, accompanied by a tidal wave of immense proportions.
                Record
Oct. 1st   9 P.M.   Bar. 30.20   wind N.E.   alt. 30
" 2nd   3 A.M.   "   29.90   "   "   " 35
"   "   7   "   "   29.60   "   "   " 40
"   "   8   "   "   29.45   "   "   " 45
"   "   9   "   "   29.35   "   "   " 50
"   "   10   "   "   29.25   "   "   " 55
"   "   11   "   "   29.12   "   "   " 57
"   "   12 m   "   28.99   "   "   " 60
"   "   12.40 P.M.   "   28.88   E.S.E. "   " 65
"   "   12.50   "   28.90   S.E. "   "   "
"   "   2.30   "   29.10   "   "   " 60
"   "   4   "   29.27   "   "   " 50
"   "   7   "   29.45   "   "   " 40
" 3rd   7 a.m.   "   30.20   "   "   " 15

The consequences, and effects of this storm were awful. — Some of which are as follows.
There were 32 persons drowned

Page from Legare's Journal on which he reports Darien's October 2, 1898 hurricane.

The tide [rose] 8 1/2 feet above the spring tide mark. It was an awful sight. My dwelling suffered very little, being probably the dryest in the neighborhood, but I had five fine shade trees uprooted.

On Champneys Island not a building remains. There is one large break, at #14, and six small ones, and the banks are badly washed all around. Of our 12 mules 10 are missing. There is not a trunk in good condition on the place. Four of my calves are missing and all my poultry. I have lost all my books and letters, tables, desks and a new carpenters tool chest that I had just finished last April at a cost of $13.40.

The following persons are now missing. On Wolf Island—A. Stokely & wife, Mrs. Cobb[138] and two little daughters & one of the Poppells—all white, and a woman on Union Island dropped her baby overboard in trying to climb a tree and it was lost. Young Jones was drowned at Wolf Island Lighthouse, making eight white persons so far drowned or missing.

This minute is written at Brunswick while on my way to Savannah to see the owners of Champneys Island and to find out what they wish done.

Oct. 6.   I saw President Comer and Acting Treasurer W. C. Askew today. Mr. Comer directs me to put Champneys Island in condition to plant another crop, to do the work as cheaply as possible but to do it well. This is very gratifying and I leave for Darien this evening.

Oct. 7.   All the mules have been located and seven have been recovered. I have also found the larger portion of my tools, but my office fixtures, bed &c. have not yet been found. Mr. Aleck Stokely, thought to have been drowned, has come in from Egg Island paddling himself to Darien in a trough. Mrs. Stokely was drowned. Carrie Magwood's body has been recovered.

The Champneys Mill is blown to pieces and lying on the upper end of Butlers Island. The Generals Island mill and other buildings are gone.

Oct. 18.   Five of our mules were found in a swamp on Elizafield in Glynn County and I had to go and get them on Sunday, Oct. 9th inst. The mules have now been recovered uninjured.

The wreckage of the Str. *City of Macon*, the Wolf Island Club house and the Wolf Island Lighthouse, and Dent's new mill are all mixed up with our Champneys wreckage every where. The Engine & Thresher are lying on their sides badly broken. Our carts are blown for miles and only seven of the 12 have so far been located.

No other bodies have been found. The water rose about 13 feet on Champneys Island at the settlement. My greatest loss is my Diary and all my memoranda[139] which I hope may yet be recovered, all my letters and a great many valuable papers are also still missing.

I find it impossible to get hands. The lumber people are paying common laborers from $1.25 to $2.50 pr. day & rations and the negroes will not therefore work for me for 75 cents pr. day and so not much has been accomplished so far.[140]

Oct. 21.   There is considerable freshet in the river. In fact, the river has been very full ever since the storm. This, and the lack of hands, have rendered it impossible for me to save my rice. There is not much in condition to be saved, but I might have saved that little. As it is I may save as much as 100 bushels in a more or less damaged condition in the settlement. It is strange that the bodies of the drowned negroes are not yet recovered.

Oct. 29. I succeeded in getting the Champney Flat out of the woods back of #2 and into the river again yesterday. It cost $23 to do the work, which is remarkably cheap. I am putting up the frame of the old kitchen dwelling in the stable lot, to be used at present as a shanty and afterward as a gear, feed & tool room.

One of the unfortunate results of the storm to be noted is that Jas. Walker has been closed out as a merchant, his store being sold yesterday under foreclosure of mortgage and Adam Strain's Sons bought it as a whole and will continue to do business.

Dec. 4. Yesterday Mr. D. S. Sinclair sent me word that he had found my desk on the upper end of Butlers Island. I went & got it and found my diary & papers in fairly good condition. I am very thankful to recover my diary.

1899

Jan. 9. The Rev. Lucius Ross Lynn,[141] who was recently called to the pastorate of our church and ordained Dec. 23rd [1898] administered his first Lord's Supper—the first we have had in one year—yesterday, his text being I Cor. 2, 23-27th verses. On the place [Champneys Island] I am plowing, clearing, ditching, banking and rebuilding the buildings carried away by the tidal wave. I planted today turnips, cabbage, peas, lettuce, mustard, sugar beets, radishes and kohl rabi in the garden on Champneys Island.

Feby. 13. Dreadful cold! The coldest day by 7° ever known here. Thermometer 9° at 9 a.m. Commenced sleeting at 9 o'clock last night. Then snowed, the wind shifting to West N. West and blowing a gale (clear day). The cold is extraordinary all over the U.S.

Feby. 16. The extent of damage done by the freeze cannot yet be determined. The heaviest loss will be to animals, many of which have died here. The weather was the coldest at Charleston, S.C. (7°) in her history. The thermometer registered 2 below zero at Tallahassie [sic], Fla. and Albany, Ga.! As I have but little to lose my loss will not be heavy. Some of my turkeys froze to death. A great many birds are killed. I have not heard of any deaths in this community from the cold.

March 6. I commenced planting today in square #1. The Negro women are at work on Champneys Island today for the first time since the storm of Oct. 2nd 1898 and I record the fact with pleasure. In spite of the rain, the mills ran all day but are giving some trouble.

March 9. I finished planting #1 & 2 today with #2 white seed at the rate of 2 1/4 bushels pr. acre and I put tonight's tide on them. It is warm enough tonight to be comfortable without fire.

March 17. Rained some Wednesday night. I turn off the sprout flow from the rice planted today. My family spent the day on the Island (wife has been here with me since Wednesday with Houstoun).

March 25. I finish planting #4, 6, 8 & 9 & put tonight's tide on them as sprout flow. Sown with #2 white rice winnowed at 2 1/8 bushels pr. acre. The rice planted hitherto was not winnowed; sown at same rate pr. acre.

April 11. Margie is laid up with the mumps. Emma has about recovered [from fever].

April 17. I turn the sprout flows off #24 & 25. A N.E. wind is blowing stiffly so

I put the stretch flow on #14, 15, 16, 17, & 23 tonight.

April 18. I lower the stretch flow on #4, 6, 8, & 9 today. Wife has the mumps. No work being done on the place.

April 25. I am planting the settlement and put tonight's tide on it. I am planting some fine Japan Rice sent out by the U.S. Dept. of Agriculture. I got two sacks from Jos. Mansfield. I am to give Mr. Mansfield a portion of the yield for his seed...The crop prospects, so far, are excellent.

April 26. A *private* tornado struck our new mill on Champneys Island—weather boarded and shingled—and blew it down, breaking our thresher. I did not hear of any dangerous wind in Darien, and therefore think the tornado must have been confined to Champneys Island. With Job I can say—"For He breaketh me with a tempest and the lessons of God do set themselves in array against me."

This is my 47th birthday and the above intelligence communicated to me this morning proved an unwelcome birthday present but, "It is well."

I turn the stretch flow off #1, 2, 3, 5, & 7 today.

Street scene, Darien ca. 1899, at courthouse looking north toward jail.

Britt & Claudie gave me a birthday supper tonight, which is much appreciated.

April 28. More bad luck. I had made the trunk bars & bed to put in a new trunk to #25, when the wind went to N.E. and blew up such a large tide that it broke my bar badly; so I had to stop the disk again at considerable expense and abandon the project for the present. #17, 18, 21, 24, & 25 were flowed but I hope no great damage was done them.

May 3. I commence hoeing regularly today in squares #1 & 3.

May 15. I opened the trunks slightly so as to turn the stretch flow off #14, 15, 16, 17 & 23 on Saturday p.m. and the water is all gone today—probably went dry yesterday.

May 16. I bought today Kingsland & Douglas "Louisiana" Thresher from H. S. Ravenel,[142] Trustee for Jas. Walker, Bankrupt, for Champneys Island, paying him $50 for it as it lies in a ditch on Rhetts Island.

May 29. The caterpillars are so bad in #9, & 16 that I flow these squares today. They are also becoming troublesome in the N.E. new ground. I have never seen one fourth as much volunteer in a crop of rice in all my experience. Area #1 from the mill downward has little in now, but it is thick every where else on the place.

May 31. It is very warm and dry. We had a sprinkle of rain last night—but not enough to show this morning. It is so dry that the Elder bushes are dying on the sides of the banks.

Thursday, July 6th. Fire in town. The following buildings were burned. The Faries[143] house occupied by Mr. Wm. Hunter and the beautiful Presbyterian Church building. Several other buildings caught but were put out. The Hunters lost the most of their furniture.

The furniture and doors & sashes of the Church were saved in a damaged condition. The Church cost $4,100.00 and was insured for $2,000.00. We are going to try to rebuild at once. This is a heavy blow to us as a congregation.

July 10. The Presbyterian congregation met at the Methodist Church on the Ridge yesterday and determined to rebuild their church just as it was. T. A. Bailey, A. E. Dimmock, W. S. Mallard & myself are the Building Committee.

July 11. The Central of Georgia Ry. has bought my Generals Island engine and boiler for $100.00 and I am taking it to Champneys Island today.

July 19. We had a very heavy rain about midnight last night. It rained again all the afternoon. After 2 o'clock lightning killed a young negro woman named McIver in Darien, and struck two trees in Reidy Walker's[144] yard.

July 20. I went fishing in Snow's Creek today but did not have much luck.

Aug. 7. I commenced harvesting today in #2 on Champneys. I have never but once commenced so early.

Aug. 18. I commence threshing today afternoon. Very hot.

Wednesday, Aug. 23rd, 1899. The negroes prevented the sheriff from taking Henry Delegal to Savannah [from Darien] this morning for safe keeping. Delegal is charged with rape of a white woman named Matilda Hope. The negroes believed that he was being taken from jail to be lynched. Thereupon, the Governor[145] ordered about 200 troops from Savannah who arrived here [by rail] about 7 p.m. and took off with them. About 100 men remained in town all night (Savannah men) with the McIntosh Dragoons. There is great excitement and trouble is feared as the negroes are quite violent in their talk. It is hoped, however, that it will end in talk. It seems that the dispatch of the troops was a salutary lesson for them.[146]

Aug. 25. There is a great commotion in the community. Messrs. O. Hopkins and Joseph E. Townsend[147] were sent out about 16 miles last night to arrest the sons of Henry Delegal, and in making the attempt the negroes shot Mr. Townsend in the abdomen, from which wound he died soon after. Mr. Hopkins was shot in the shoulder, his being only a slight wound. Mr. Hopkins shot at one of the negroes and although wounded, succeeded in bringing Mr. Townsend away with him. There is intense excitement in Darien and a lynching bee for tonight is probable. The authorities have arrested & jailed so far about 35 of the more prominent negroes, both men & women. The Superior Court will meet in extra session next Wednesday to try Delegal & the others. The trouble is by no means yet over. The Savannah troops are still here.

Aug. 26. Rained quite heavily last night. Today a large number of men, including the Liberty Troop, are out in the northern section of the county on a regular man-hunt for the murderer of Mr. Townsend. It is believed that the negroes are out there in force and that there will be some fighting. Mr. Wm. R. Townsend is the leading spirit. He has blood hounds with him.

Aug. 30. The Superior Court met in extra session today to try the participants in the late race troubles and I am elected Foreman of the Grand Jury for the 7th term.

Soldiers are still here. Col. Lawton and the companies of the 1st Georgia Infantry in Savannah are here until tonight when they will leave for Savannah and they will be replaced by four companies of the 1st Georgia Cavalry, Col. Lawton being still in command. These troops are the Wayne Troop, Liberty Independent Troop, Liberty Guards and the McIntosh Dragoons.

Friday, Sept. 1. The Grand Jury has taken a recess until tomorrow morning having found true bills against 39 negroes for rioting and three negroes for murder, and no bills against 22 darkies for lack of evidence. The negroe rioters were taken to Savannah for safe keeping but were returned here for trial yesterday.

Sept. 2. I shipped today by tug *Maggie*, Capt. Judkins, 1680 bushels of rough rice[148] to Savannah.

Sept. 7. The extra session of the Superior Court adjourned today. There were about 21 rioters convicted and fined—the fines aggregating $10,500.00. Henry Delegal charged with rape—the cause of the whole trouble—was tried and a mistrial had. The Judge had previously granted a change of venue in the case of the State vs. John, Edward and Mirander Delegal—murder—he today granted a change of venue to Henry also. All the prisoners will be tried in Effingham County next week.

The troops left Darien today for their respective homes. The cases against the rioters have been appealed and the most of them are out on bail.

Sept. 9. I shipped by Capt. Judkins tug *Maggie* today one lighter full, 1667 bushels rice by weight. This rice turned out 1703 bushels in Savannah.

Sept. 16. We hear today from Effingham County that Henry Delegal has been acquitted of the charge of rape, and John Delegal charged with murder has been sentenced to the Penitentiary for life.

Sept. 18. Edward & Mirranda Delegal were tried in Effingham Co. last Saturday for murder of Jos. E. Townsend. Edward was sent to the Penitentiary for life and Mirranda was acquitted.

Shipped today by Capt. Judkins one Lighter full—2340 bushels white rice by weight. Total shipped to date. 5723 bushels.

Sept. 23. Shipped today by schooner *Oakland*, Capt. John Col[e]man, 2492 bushels rough rice. Total to date. 8215 bushels.

Oct. 3. Shipped today by schooner *Oakland* 2500 bushels rice in bulk and 73 sacks, 355 bushels rice in sacks. Total shipped to date. 10,971 bushels.

Oct. 12. Shipped today by Capt. Judkins one Flat full, about 1700 bushels rice. Total shipped to date. 12,738 bushels.

Oct. 14. Shipped today by schooner *Oakland* 2452 bushels rice, being full except the fore hatch which required about 50 bushels to fill it. Total shipped to date. 15,090 bushels.

Oct. 18. On a trial of my several boats today the following is the speed record for one mile, 1/3 mile with the ebb and 2/3 against the ebb tide on the same tide over the same course and by the same oarsmen. "Emma", one mile 17 min. 30 sec. "Lizzard" one mile 18 min. 0 sec. "Eel" one mile 16 min. 35 sec. "Varuna" one mile 16 min. 30 sec.

Oct. 19. The mill has done nothing since midday Tuesday when the thresher broke. Two of our flues were leaking also. I have repaired the thresher but find that the flues are in such bad condition that I cannot plug them. I will have to put in two new tubes, which I have sent to Brunswick for.

Oct. 25. I shipped yesterday pr. schooner *Oakland* 2376 bushels rice being her hold full. Shipped to date. 17,466 bushels.

Oct. 28. I got through threshing all this year's rice except after rice today.

Crop 1899. Total for the crop 19,605 bushels from 410 acres net or 47 3/4 bushels pr. acre. Straw account. Delivered J. A. Britt 190 bales, 211 bales, 223 bales, 215 bales. Sold J. H. Judkins 40 bales.

Nov. 13. Britt & Claudie had their second daughter [born August 24, 1899]—Emily Wilkinson—baptised at Sunday School yesterday afternoon by the Rev. L. R. Lynn.

Dec. 12. All last week I hauled shell and sand for the [rebuilding] Presbyterian Church with Champneys mules, carts and men. But we are all back on the place today. Work on the Presbyterian Church was regularly commenced—the foundations being dug—on Monday, Dec. 11th 1899. We hope to get it finished by July 6th the anniversary of the fire.

1900

Jan. 17. There is much commotion in the community owing to the fact that several cases of small pox have been discovered. It is also prevalent at places widely separated in the southern states, and it is in Brunswick, Waycross, Jesup, Johnstons[149] and at several other places near us.

Jan. 22. Charles O. S. Mallard[150] was ordained an Elder of our church yesterday.

Jan. 26. I am having piling driven by Rolland Jackson at $7.50 each. Four at the mill wharf, two at #15 trunk, three at #24 trunk, two at #26 trunk, in order to properly bulkhead the trunks on the outside and keep them from "blowing out."

Feby. 28. The wind blew hard from N.E. all last night until about 10 a.m. when it shifted to S.E. The high N.E. wind backed the freshet up so that this morning's tide flowed up the place so that I had to stop planting. I will put tonight's tide on #1 which is finished and #2 which is about half finished and will plant the rest of #2 later, working the middle water on it. Squares #4, 6, 8, 9, 12, 13, 18, 19, 20, 21, 22, 24, 25, 26, 27, 28 and 29 are flowed up today. #2, 3, 5 & 7 are badly wet. The settlement is dry so far and I hope will continue so.

March 3. I plant my cane today. The freshet is leaving the place. (The settlement has not been flowed).

March 7. Some one tried to rob the mill by boring three 2-inch auger holes through the weather boarding, but they only got about 1/2 bushel of rice.

March 10. I planted square #1 on Feby. 27th and had gotten about half of #2 planted when the strong N.E. wind blew the tide over it and flowed it. Today I find these two squares sprouted so I dry them.

March 12. I am planting #3 today. I hope to get #3, 4, 5, 6, 7, 8, & 9 and the rest of #2 on these tides. I have until the 20th inst. in which time I may get water to flow them.

March 17. #8 is planted with J. T. Dent's Gold seed at the rate of 2 1/4 bushels pr. acre. The other squares are sown with our own white seed at the same rate.

March 29. I plant the balance of #2, about 15 acres, today and put tonight's tide on.

April 6. Still too cool for the season, so cool in fact, that I am afraid to keep the water on the second planting (flowed two days ago) for fear of drowning out the low places, so I dry them today. This is an experiment, but I hope it will work out all right in the end. The rice is so small that I cannot trail it.

April 11. I plant the settlement with our own white seed and put tonight's tide on it. I put the stretch flow on #3, 4, 5, 6, 7, 8, & 9, also tonight. The stand seems to be fairly good in #1, 2, 3, 4, 5, 6, 7, 8, & 9.

April 14. I finish planting #13 today and put tonight's tide on it as sprout flow. It is sown as follows: Beds 2 to 12 = 13 acres, are sown with Japan rice raised in the settlement last year. The rest of the square is sown with Gold seed bought from J. T. Dent, all at the rate of two bushels pr. acre. The southern ends of beds 13 & 14 are also planted with Japan rice, making about 13 acres planted in this rice.

April 27. I turn the stretch flow off #1 today. The tides are immense and the high river is causing trouble by flowing the low places in the upper squares.

May 2. The Episcopal Rector, the Rev. Mr. Woodward, shot a negro man named John Jones last night, giving him a dangerous wound in the abdomen. The negro worked for me at times. The shooting seems to have been justifiable, but there are some things about it that are inexplicable—to me, so far, I have heard both sides but some things are unexplained.

May 4. John Jones died yesterday evening from [his] wound. The Coroner's Jury pronounced it "justifiable homocide."

May 11. The Sundays Schools of the county (except the Episcopalians) are having a picnic at Crescent today. None of my family will go.

May 18. I am planting ash & gum on Champneys Island. Today the most seasonably warm day so far this season. I have a large amount of rice suffering for hoeing and I cannot get hands to hoe it.

June 4. I put the stretch flow on #27 and 28 tonight. Hands are still hard to get and I have 250 acres in need of hoeing, of which 140 acres, namely squares #4, 5, 6, 7, 8, 9, 12, 13, & 19 are suffering for work. I am afraid that some of it may have to be abandoned if I do not get a large number of hands soon.

June 5. I have 31 hands today and so the crop outline is brighter. I have just bought, and am using, four rice cultivators made in accordance with my ideas, but made & patented by W. S. Mallard of this place. They are doing fine work and I am much pleased with them.

June 13. I was elected an Academy Commissioner by the last Grand Jury. These Commissioners held their first meeting today and I am elected Chairman of the Board which will meet the 2nd Wednesday in June, September, December & March of each year.

June 20. The Railroads are giving cheap rate excursions to Savannah today, and a large number of persons from Darien are going. Britt, Claudie, Margie & Emma have gone. They will return tonight. I put the harvest water on #2. Someone tried to burn the Depot[151] in Darien last night, but the fire was discovered before much harm was done.

June 23. The following places seem to be suffering for work. The hills in #1, 2, 3, 5, & 7. The rice is losing its color from lack of sunshine. On the whole, the crop is a good one, but I do not think that the hills and very low places any where will do much. My experiment of putting the harvest water on the lowest parts of # 3, & 5 is a success.

June 27. We are informed that the rivers in the upper part of the state are all very high, so I confidently expect a freshet of considerable size. The head waters will be due here July 4th but the worst of it will be here July 9th. I expect trouble. I fear [T. H.] Gignilliat & C. O. S. Mallard will be drowned out entirely.[152] Old Mrs. Cromley[153] died this afternoon of cancer of the stomach on the Ridge.

July 4. The Confederate Veterans had a picnic at Eulonia .

July 6. Today is the anniversary of the loss of our church building by fire. We are rebuilding and the new church is rapidly nearing completion. We hope to occupy it Sept. 1st. It is built of Tabby concrete[154] with a slate roof. It is a handsome structure. The wood work is about finished, except putting in the sashes and doors. Plastering is commenced today. The tinning is nearly finished and painting is to be commenced next Monday.[155] Today is the hottest day so far. Temp. 93º in the entry way to the Champneys Island house. I put a thermometer in the sun with a white linen handkerchief on it and it registered 10 minutes later 108º. I then put the thermometer in the sleeve of my black alpaca coat and exposed it in the same way for 10 minutes and it registered 116º–a difference of 8º on account of the difference in color.

July 12. The height of the expected freshet ought to be, and probably is, here today, but it is doing no harm. The lower part of #28–about three acres–is flowed and that is all. The last July freshet occurred in 1886 but that freshet was *much* higher than this one.

July 30. Arthur Hamilton[156] was shot by an assassin in Darien on Saturday night, about 10 o'clock. He was shot with buckshot from a gun, the load lodging in the right iliac region about six inches from his backbone, four of the shot coming through the front. The lead entered the abdominal cavity. He was taken to my house where he died yesterday at 12.47 p.m. We buried him in St. Andrews Cemetery this morning on a lot I purchased there. There is considerable excitement in the community over the affair as it is thought to have been the work of a "Striker." The negro Longshoremen at the Sapelo loading grounds[157] have struck for higher wages and shorter hours, which the timber merchants will not allow.[158] And Arthur had [illegible] to work there. A huge reward will be paid by the city, county and Lumber merchants for the arrest of the murderer. Wife is about laid up over it. As yet there is no clue to the perpetrator.

Several negroes have also been shot in the community lately, apparently for the same reason.[159] It seems that there is an organized plot to shoot all who take the places of the strikers.

Aug. 6. The authorities arrested Smith King—my foreman—Saturday and took him to Savannah with several other negroes charged with complicity in the murder of Arthur Hamilton. Smith was with Arthur when he was shot, but neither he or Arthur could tell who did the shooting. Smith is not thought to be guilty, but the authorities hope to get some information from him that they think he could give if he would. My opinion is that Smith has told all he knows. I am promised that he will be released about Wednesday next.

Aug. 20. So far Smith King has not been released, nor will the commitment trial be held until the 28th of August. We have reason to believe that at least one of the guilty parties has been jailed, but the others [are]—we fear—still at large.

Aug. 28. After keeping Smith King in jail for 24 days without a hearing he was released today—being discharged from custody as a suspected murderer—but held under a $100 bail bond as a witness in the case of the State vs. Tom Elverson & Richard West charged with the murder of Arthur Hamilton.

Sept. 1st, 1900. The Presbyterian Church building is now finished—except a few touches of paint—and we hope to dedicate it to God tomorrow morning. The Rev. Albert B. Curry, D. D., twice Pastor of this church, but now of Birmingham, Alabama, will conduct the service, assisted by Pastor Lynn. Mr. Curry will spend today and tonight at our house.

Sept. 3. The Presbyterian Church was successfully dedicated yesterday in the presence of a large congregation composed of the members of the different white congregations of Darien who attended by invitation.[160]

Rained some in Darien yesterday morning, and again yesterday evening, and again today, but very little rain has fallen on Champneys Island. Today I bought and paid $20 for Lot 82 in St. Andrews Cemetery. It will be owned by Britt & myself jointly.

Sept. 5. Last Monday at the preliminary trial, Tom Elverson was held as the murderer of Arthur Hamilton—I hope it will transpire that the authorities have the right man. I commenced threshing yesterday p.m.

Sept. 10. We are just informed that the cyclone that was predicted for last Thursday, wrecked Galveston, Texas and the surrounding country on the 8th and 9th inst. killing a large number of people and doing a large amount of damage.

Sept. 11. Rice birds are more numerous and troublesome than I have ever seen them on this river. We get from 40 to 50 dozen a day and cannot keep them down

Sept. 18. I make my first shipment of rice today by Schr. *Oakland*, Capt. John Coleman, to Jno. Screven, Savannah.

Oct. 3. I am one of the Freeholders of the election held for state & county officers today.

Oct. 5. Britt has been elected Ordinary of McIntosh County.

Oct. 11. I served as a "Freeholder" at the election held Oct. 3rd. Now Mr. S. A.

Darien Presbyterian Church as it looked soon after the rebuilding was completed in 1900.

Way, the defeated candidate for Clerk of Court, is contesting the election because I am not an owner of real estate. I have no *papers*, but I do own an undivided half interest in a lot just back of our place at Ashantilly, and purchased from Mrs. B. Britt two years ago, the titles being made in Claudie Britts name, with the understanding that she is to make my titles when called for.

I also bought St. Andrews Cemetery Lot #82 for $20 about a month ago, but have received no titles yet, but *they* have the money and *I* have the lot. I also own 1/4 interest in the Ashantilly School House, but again I have no papers. The lawyers—four of them—say I *am* a freeholder. The case will be heard in Darien Oct. 19th, 1900. It is giving me much trouble, but no uneasiness, as my conscience is clear. I was asked to serve by two county commissioners and two candidates.

Oct. 20. The election contest between S. A. Way & A. C. Wylly[161] is giving me great trouble. The hearing of evidence was held at the Court House yesterday as [to] my being a Freeholder. Even Claudie was called as a witness.

It is all for nothing. Mr. A. C. Wylly has been elected and will get the office. Mr.

Way—I do not think—stands any chance of getting the office and yet he persists in giving trouble and at the same time killing himself politically. I do not feel very kindly towards him.

Oct. 26. I have had a greater amount of stealing of rice by the hands this week than in all my experience before. One negro told on others causing us to find about 20 bushels of rice that had been stolen right before our faces, while a trusted negro took about 15 bushels that has not yet been discovered. The foreman and myself are very much discouraged over it as the whole thing is the stickiest piece of villainy we have met up with of late years.

Dec. 1. Court has adjourned. Late this week we discovered the whereabouts of Charly Houston (Col) who is believed to be the murderer of Arthur Hamilton, and we have very good evidence to convict, but we desire more time in which to gather evidence enough to assure conviction; so we have asked for and had the trial of the case postponed until the May term of court in 1901. The Grand Jury found a true bill against Charly Houston and Tom Elverson at this term for the murder of Arthur.

Dec. 8. I have orders from Mr. Cunningham directing me to go to work preparing for the crop of 1901.

Dec. 27. Christmas Day was fair and very pleasant [and] passed off quietly. All at home got presents, Mr. Cunningham sending me a turkey, dressed, which was highly appreciated. No work has been done so far this week on the place.

1901

Jan. 29. My friend Albert E. Dimmock of P. S. Clarke[162] & Co. of Darien has sold his half interest in that business to Adam Strain's Sons and it is his intention to move away from Darien, which I greatly regret. He will be missed by the community and will be especially missed by, and in, our church of which he is an Elder and a pillar.

Feby 14. On the 5th inst. the Board of County Commissioners elected me a member of that board to serve out the expired term of A. E. Dimmock, resigned. Today, I received my commission from Gov. Candler and took the oath of office before Ordinary J. A. Britt. I am now County Commissioner & Academy Commissioner for the county of McIntosh; and Pilot Commissioner for the City of Darien.[163] Neither office pays any salary. The work is all to be done free—gratis, and for nothing. I will therefore not fatten at the public crib. Not many thanks even attached to either office, while the labor is considerable. But some one has to do the work, and the time has come when I am called to do my part, which will be done to the best of my ability.

March 3. **Fire!** Fire was discovered in the store of F. Nanias[164] in Darien at 3.30 this morning. It proved to be the worst since the War. The following stores were entirely consumed—

F. Nanias stock and store; I. Joselove[165] stock and store; M. Bluestein[166] stock and store; P. S. Clark & Co. stock and store; Dr. P. S. Clark instruments & office fixtures.

It is not certainly known how the fire started, but my theory is that old Mr. Nanias upset a lamp in going upstairs. He did not seem anxious to have the fire put out and the people are much incensed at his behavior. Darien looks forlorn indeed today.

March 4. I finished planting #14, 15, & 23 today and put tonight's tide on them.

They are all planted with Gold seed raised on the place from seed bought from J. T. Dent last year, at the rate of 2 1/8 bushels pr. acre.

March 5. I finished planting #16 & 17 today. Birds are very troublesome.

April 13. The expected freshet is here. The low places in squares #24, 25, 26, 27, 28, & 29 are under water.

April 16. The freshet has covered squares #24 & 25, so I consider the stretch flow on them, but I am quite uneasy about this. I fear the stand will be poor, the freshet having taken them before they [sprouts] were up.

April 18. I plant the settlement in white seed rice today, but will put tomorrow morning's tide on it. The freshet tide last night was very high and flowed the place except the settlement and squares #1, 2, 3, 5, & 7.

April 19. A S.E. gale blew all last night and this morning. This, with the freshet and high tides, have flowed Champneys (and all other places in the delta). The damage cannot be estimated today, but will be quite heavy.

April 22. The water has not yet left the upper squares on Champneys. The rice is puny, and altogether the crop outlook is unusually blue. I cannot yet tell the amount of damage done by the late over flow, but the following are some of the damages done—

I had five turkeys drowned. My garden is injured. The banks are badly washed all around and there are 8 small breaks on the place, none of which are likely to give much trouble in repairing them. I think the rice in #24 & 25 will have to be re-planted and perhaps some other places also.

April 25. The freshet water has now left the place, but the land is still very wet. I have now stopped all the breaks from the gale of the 19th inst. but the banks are so badly washed that I fear I cannot get them repaired again this season.

Another freshet is expected here about May 7th. I am afraid to do any planting. I shall therefore await developments.

April 26. This is my 49th birthday; and for the first time I must say that age is [illegible] on me. During the past year sick headaches have afflicted me and they have so far injured my sight that it is difficult to read at all without glasses. Otherwise, I am "still in the ring."

May 2. The rice is very thin in places and in some low places I cannot trail it even now. I cannot re-plant now on account of May birds.

May 4. It has been quite warm for several days and the crop has improved greatly, but still there is damage apparent and it may be several weeks before I can tell the exact amount of damage. A great deal of the rice in the low places has died already. Whether enough is dead to injure the stand will have to be determined later; but this much is certain: the loss will not be nearly so great as it appeared a week ago.

May 8. Jacksonville has had a most disastrous fire. It occurred last Friday, the 3rd inst. 136 blocks were burned entailing a property loss of $10,600,000 and making about 10,000 people homeless.

June 1. Chas. Houston, charged with the killing of Arthur Hamilton, was acquitted last night. We have therefore entered a *nol prosiqui* in the case against Thomas Elverson charged with the same offense.

June 5. Caterpillars are doing some damage to the rice. The crop has improved greatly of late and is now quite a good one. Finding the caterpillars are damaging those squares I flow #24 & 25 tonight—or will try to flow them.

June 15. I finish planting #28 with Japan seed and put the water on it. I still have #22 & 29 = 31 acres, plowed and ditched but I cannot spare the time from the growing rice to plant them. I will therefore stop with what I have planted, which makes 380 acres net planted so far. Caterpillars are eating the rice clean off the hills in squares 11, 12, 19, & 25. I flowed 24 & 25, but the flowing seemed to do more harm than the caterpillars. The crop is good. Labor is scarce and the frequent, heavy rains interfere greatly with my work, but the rice is growing.

June 24. Memorandum—The following property on Champneys Island is mine—
Water mill. The Huller & some tools in storeroom. The Fan.
Flat Dock—Corn mill, Straw Press, Rice Polisher and remains of Acme harrows.
Dwelling. 1 Forge (the new one ought to be mine because mine was worn out in the service of Champneys Island). Everything in the Dwelling except some of the tools and the bolts. Leather &c. in my work room, and the supplies in the locked room and some of the Oars upstairs.
Also the following row boat: The "Eel" with 7 prs. oars and row locks.

I will hereafter try to keep the plantations tools separate from mine.

July 2. I took my seat as Commissioner for McIntosh County today, having been re-elected by the last Grand Jury to that office.

July 3. Mrs. R. Austin Young died last night at her home in Darien after a protracted illness from a complication of diseases, leaving a husband—our Receiver of Tax Returns; and four small children, the youngest being a seven months child born about a week ago. Very, very sad!

July 20. There was a small fire in town at 3 o'clock this morning—a negro house just north of the Ice factory. This fire threatened the negro Baptist church and all the negro bells in town were rung, which made us think that all Darien was on fire. The negro house was entirely consumed.

July 30. The Board of Academy Commissioners elected the following teachers today. Principal, Prof. C. E. Cook, and Mrs. Florence Lachlison[167] and Miss Willie Mallard as assistant in Darien and Misses Annie Knox and Roberta Cook as the teachers in the academy on the Ridge.

Aug. 26. I commenced harvesting today on square 10 on Champneys Island.

Aug. 31. The weather has been cloudy-showery for several days and has interfered greatly with the rice harvest. I have more volunteer rice even on rested land than I ever saw before—square #14 (Champneys) was planted in gold rice but suffered greatly in the spring from cold and freshet. Now it seems to have been planted with white seed. The volunteer is the green cuticled variety and I hope to save it, but it is shattering off badly now although not quite ripe enough to cut.

Sept. 7. We have just received the shocking news that Maj. Wm. McKinley, President of the U.S., was shot and is thought mortally wounded by an assassin with the ever ready pistol in Buffalo, N.Y. His assailant is an anarchist named Czolgozs [sic]. It is hoped that he is not fatally injured. Mr. McKinley is one of the most popular men who ever occupied the presidential chair.

Sept. 14. Pres. McKinley died in Buffalo, N.Y. at 2.15 this morning from the wound inflicted on the 6th inst. He is universally mourned. Vice Pres. [Theodore] Roosevelt has taken the oath and is now President of the U.S.

Sept. 21. I make my first shipment of rice today by Capt. J. H. Judkins being 2128 bushels white rice from #10, & 18 and the settlement—50 acres, making the yield 42 1/2 bushels pr. acre so far. Rice birds are apparently leaving.

Sept. 26. The wind has blown hard from N.E. for about a week, at the rate of about 35 miles an hour, and the tides are immense. #4, 6, 8, & 9 on Champneys, and most of the other places are flowed more or less. No damage has actually been done me yet but tonight's tide promises to be very high.

I have about 1700 bushels of rice threshed and about 400 more unthreshed in the mill.

Sept. 27. Champneys Island is flowed, on an average 2 ft. deep, but the stacks are not yet floating. The banks are badly washed in places, especially the south bank to the place, but there are as yet no breaks. The loss will be considerable even if no worse damage occurs.

Sept. 28. I saw F. H. McFarland,[168] President of the Darien & Western Railway, on the street this morning, so drunk he could hardly stand! What are we coming to!! The wind has changed to S.E. and the weather has faired off, but the tides are still immense and Champneys Island, and in fact, the whole delta, is flowed up even deeper than it was yesterday.

Sept. 29. I find things on the place no worse than they were yesterday. The settlement is now dry and the N.E. new ground, but the rest of the place is still flowed.

Our Pastor, The Rev. Lucius Ross Lynn, resigned the pastorate of our church today and his resignation was accepted; because our church has been lately so impoverished by removals, that we are not at present able to support a Pastor. He has made us a most excellent Pastor and we part with him with deep sorrow.

Oct. 4. The height of the freshet ought to be here today. The upper squares are flowed to about the upper gauges—(8 in. deep). This greatly hampers me in drying our wet rice. Our loss will probably reach, and perhaps exceed, $4,000.00

Oct. 10. Mr. Lynn gave me his bicycle. It is Fetherstone Model A. #818. 30 in. wheels, G.&J. tires, with 20th Century Lamp, bell & pants guards. My Bicycle is Hawthorne 1900 Model #280, G.&J. tires, gear 72. The gear of Mr. Lynn's bicycle is 80.

Oct. 11. The negroes have had the Champneys Flat from Sept. 23rd until today—they got it on an order from the Hilton & Dodge Lumber Co. I charge them $1.00 per day. They had it 15 days. The H. & D. L. Co. therefore owes me $15.00 for which I will send them a bill today.

Oct. 15. Statement. Shipped Sept. 30th 1646 bushels. Shipped previously 2128 bushels. Shipped to date 3774 bushels. This is all good, undamaged rice.

Nov. 30. The Rev. N. Keff Smith, D.D.[169] arrived here tonight for a visit and is staying with me. Our church has called him again as Pastor and we hope he will accept the call.

Dec. 14. Lectured Crawford,[170] the colored representative from this county in the [state] legislature, died in Atlanta two days ago. His body was brought here for burial yesterday. The negro Masons are standing guard over his remains with drawn swords

& all such foolishness, and the brass band is serenading the corpse.

Dec. 31.   Rev. N. Keff Smith, Mrs. Smith and their children arrived last night and are staying with me at present—until their household furniture arrives when they will stay in the Wilcox tabby building at Ashantilly. Mr. Smith has accepted the pastorate of our church.

1902

Jan. 2.   My first grand son [Legare Britt] was born to Jesse Britt and Claudie at 5 o'clock p.m. today.

Jan. 15.   F. G. Parnell's only remaining son, Alva, died last night of scarlet fever, making two children lost in 48 hours and their only remaining child, a daughter, is now very ill with the same disease. A great many people, myself among others, visited the house before it was known that the children had scarlet fever and there is much uneasiness in the community because of that fact. The schools are closed for the present.

Jan. 20.   Frank Parnell's only remaining child, a daughter, died of scarlet fever last night. This makes the third, and last, child he has lost in a week! N. H. Barnwell paid me a visit today on Champneys Island.

Feby. 12.   The Darien & Western Ry. had two serious accidents today. They ran a special train which killed an old negro woman—the mother of Ned who works with Britt. The old woman is entirely deaf, and was attempting to walk from Darien to her home at Eulonia when she was run over at or near Rogers Shop, a short distance from Darien. Then, the same train ran into a box car at Eulonia and knocked it to pieces and damaged the engine considerably also.

Feby. 28.   Smith King overloaded a boat today with clean rice and sunk it getting five barrels of clean rice wet! What will be the loss I do not know, but I fear it will be quite heavy.

March 3.   Yesterday morning I received notice from the U.S. Weather Bureau in Macon that a very large freshet is coming down, the largest since 1887. I am doing all I can to keep it off the place.

River Information—The highest water recorded at Macon, Ga. was 26 feet, 9 inches in August 1887. This is commonly called the "Cleveland freshet." It is about 32 inches higher at Macon than the "Harrison Freshet" in 1847. The danger line at Macon is 18 feet.

March 6.   I commence planting in #1 today. This square is sown with Japan rice at the rate of 2 1/8 bushels pr. acre. I will not put the water on it at once but will wait until I see what the expected huge freshet will do.

March 18.   The freshet is falling. It is not as large as we expected—in fact, I have seen several larger freshets in my 25 years residence here. No damage was done. I am laid up with headache all day, Sunday, and had it quite badly all Saturday and Monday.

March 22.   The freshet keeps the river so high that what little water is in the lower place cannot run off.

March 25.   No work can be done because the river is so high that the water cannot leave the place. The water is actually increasing from rain & seepage. I received two tons of ground fish fertilizer today, which I will experiment with. I will put most of it in #17.

March 31. The freshet is leaving the place. But we received notice this morning that another freshet is on its way down. So, I must hurry the rest of the crop into the ground.

May 26. My little grand son, Legare Britt, was baptised yesterday by Rev. N. Keff Smith.

June 4. Emma graduates from the high school of the McIntosh Academy today. Houstoun is advanced from the fourth to the fifth grade in the Grammar school of the same Academy.

June 6. I left Darien today for Cross Hill, S.C. I arrived at Columbia at 7 o'clock p.m. and was met there by brother in law Herbert T. Thomas who took me to his house to stay while in the city. There I met Mother, Bro. Frank, Sisters Sue and Edith Thomas and her children. They treated me as well as I could wish and so I had a pleasant visit.

June 7. I left Columbia at midday...arrived at Cross Hill at 3 p.m. I found things somewhat better than I expected.

June 10. I have met a great many of my old friends and some new ones [and] am treated with great kindness by all.

June 16. I got home yesterday evening at 6 p.m. I visited the Island today and find things in good shape—except that the caterpillars are doing injury still in square #23. Squares #14, 15, 16, 17, & 23 have been flowed for 48 hours each since I left here a week ago.

July 16. Charles Reide Walker, our next door neighbor, married his deceased wife's sister, Miss Annie Richardson, at Defuniak Springs, Fla. today. He and his bride are expected home tomorrow on the *Hessie.*

Aug. 21. I am on the Island today where I find everything working smoothly. The rice harvest was commenced in square #2. The heat yesterday and today [was] almost unbearable. The thermometer registered 98º this afternoon. There is almost no breeze.

Sept. 18. The hands on the place are giving me trouble. They don't want to work at harvesting because they say Mr. Dent is paying $4.00 pr. acre for cutting fallen in rice, which of course is a falsehood. I have only 5 hands cutting in #1 at midday. The settlement has fallen down badly and is sprouting, but I cut #1 first so as to prevent its getting in the same fix.

Sept. 23. The weather has faired off and I am doing all I can to save the crop but the negroes are not working as they ought, and are not as plentiful as I would wish. I find that the late rains have damaged me more than I have hitherto thought. Some of the rice is badly sprouted and the bottoms of the stalks every where are wet. I fear that nearly all of the rice cut before the rains will have to be over hauled and restacked before it can be threshed.

Oct. 4. I shipped today by Capt. Judkins lighter about 2100 bushels of white rice of which about 625 bushels was Honduras rice weighing 47 lbs. pr. bushel. The other rice weighed 46 lbs. pr. bushel on our scales.

Oct. 18. Statement. Shipped last night 2227 bushels. Shipped previously 1999 bushels. Total shipped 4226 bushels.

Oct. 27. Rained all Saturday, after 10 a.m., all Saturday night and all Sunday. Fully

Hunter, Benn & Company timber boom.

12 inches of water has fallen and the lower places every where in the high lands are flooded. The Shell road cannot be travelled either on foot or on Bicycles because the water is half knee deep at the culvert and at the mile post.

This is the heaviest rain I ever saw. The Darien & Western train is at Hudson[171] between two washes and can neither come in nor go out.

On the plantation the squares were flowed up—the upper ones over a foot deep. The lower squares about midway between the gauges (from 4 to 6 inches deep).

Oct. 29. Sundays, Mondays, Tuesdays & today's mail came in today via Townsend S.A.L. Ry.[172]

Oct. 30. The Darien & Western Railway is opened again for travel today after having been closed since Sunday evening [October 26].

Nov. 7. Statement. Shipped by Capt. Judkins today 2306 bushels. Shipped previously 4226 bushels. Shipped to date for [1902] crop 6532 bushels.

Nov. 20. Shipped today by Capt. Judkins 1100 bushels more or less Japan Rice from #1 & 16, and about 500 bushels of slightly sprouted rice. Statement. Shipped today 1563 bushels. Shipped previously 6532 bushels. Shipped to date 1902 crop 8095 bushels.

Nov. 28. We had a high S.W. wind last Thursday that interfered greatly with my bringing a flat load of wood from Darien[173] and made it cost about twice as much to bring it as it ought to have cost. We ran the mill until 4 p.m.

Dec. 1. I am taking our boiler to Darien for Aleck Bailey to overhaul. I am going to bring his "Donkey" boiler over and try it, as he will be a month repairing ours.

Dec. 10. Received orders from Mr. Cunningham today to commence work as soon as convenient in preparing land for another crop.

Dec. 19. I made a trip to Savannah, leaving home on Wednesday, and getting back Thursday night. I went to buy some necessary supplies for my family. I took Houstoun with me in order to have his teeth fixed.

Dec. 27. I was taken down with the Grippe and had to go to bed Wednesday and stayed there Thursday and had to keep in the house Friday so that I am out today on the Island for the first time since Wednesday morning. Otherwise, we have had a pleasant, merry Christmas and we have much to be thankful for. I am not well yet, but am much better.

Dec. 31. Sam Haywood, the mill watchman, came over drunk on Monday night; made considerable disturbance on the place, first by cursing and shooting at Charles Powell, the shot making large holes through the house he was in. He then threatened to shoot the foreman, Smith King. I had Sam arrested yesterday on two warrants. The case was called this morning for the preliminary trial and, as the witnesses Charles Powell, Ret. Brown and Tom Grant seem inclined to lie out of it, I had the case adjourned until 5 p.m. today at which time I will make Smith King testify. But it is my present wish to abandon both cases and fine Sam Haywood heavily for his misconduct. I have $23 of his money in hand.

1903

Jan. 8. Fire in town—Mr. Robert Manson's[174] house was burnt about midday today.

Jan. 9. I received the following articles from Savannah for Champneys Island today: 20 sacks oats, 6 Perry plows, 1 coil rope, 3 canvass collars, 1 doz. B. B. hooks, 1/2 doz. shovels, 1 keg 10d nails, 1 side harness leather, 2 sheets zinc.

Jan. 22. I shipped today by Capt. Judkins 1 Lot of white rice in good order. Statement. Shipped today 950 bushels. Shipped previously 8095 bushels. On hand scraps about 265 bushels, after rice 1000 bushels, seed 800 bushels. Total for the crop is 11110 bushels.

A negro who lived near us on the Shell road named Powell Mongin, or Williams, I am not certain which, was killed at the Lower Bluff Mill today in trying to put a belt on a moving pulley.

Feb. 28. Emma and little grand-daughter Charlotte Britt are down with the measles. Houstoun is out again, having fully recovered from his spell of measles. I think the other two grandchildren are also taking it.

March 6. I went to Savannah Wednesday to buy 2 mules. I bought a pair from [illegible] & Sipple for $245.00 I took supper with Isidore Collat[175] & family and hired a room at the Planters Hotel. I got to bed about 11 p.m. At 12.30 the hotel boy awoke me with the cry of "fire!" at my door. I dressed quickly and found that the fire was in the grain store of W. D. Simpkins just across Bay St. Lane from the Planters Hotel. At the fire I heard that there had been a disastrous wreck on the Seaboard Air Line Ry. I got the clerk of the hotel to telephone the train dispatcher and find out whether the wreck would probably interfere with my return by that route. He said the S. A. L. trains would go to Jacksonville over the Atlantic Coast Line's track. I then asked for information at the Union Depot ticket office. They did not know anything about it, so I thought best to try to reach home via the Atlantic Coast Line and Brunswick. I therefore left Savannah at 4.30 a.m., arriving at Darien by the steamer *Hessie* at 1.30 p.m. I was well worn out and slept the rest of the day.

Emma and little Charlotte have about recovered from the measles. Little Emily has it broken out on her now. The mules I bought in Savannah will be brought down by road by Charlie Fulton. I hope they will reach here tomorrow.

March 21. Mr. R. E. Briesenick of Brunswick notified me by telephone that the new boiler ordered for Champneys had arrived in Brunswick in a badly damaged condition, it having been in a railroad wreck. He also told me that he had a good second hand boiler for sale cheap, so I concluded to go and see it. After consulting Mr. Cunningham I went over Thursday morning [to Brunswick] on the *Hessie*, taking Houstoun and little Charlotte Britt with me. By his pressing invitation we stayed at Mr. Briesenicks house. He, Mrs. B. and the children treated us as kindly as they could have treated near relatives. Friday morning Mr. B. hired a carriage and took us to see the new U.S. Custom House which he is building, then to see the boiler, then to the machine shops and then to the *Hessie*, which we took for Darien arriving at home to dinner. The children enjoyed the trip greatly. I would have also, but I had a headache all Friday.

The Boiler is a 25 horse power return tubular one, in good condition. He will finish it and the necessary outfit for $150.00. I like it very much, so after consultation with Mr. Cunningham & the purchasing agent, I have taken it.

April 27. Mrs. Lillian—wife of William W. Sinclair—died suddenly, it is supposed of heart trouble, Saturday night. She leaves 5 small children. Sad! It is not quite a year since Mrs. Emma—wife of the elder brother, David S. Sinclair—died in very much the same way. Yesterday was my 51st birthday. I spent it in bed with a sick headache.

May 1. I finish planting #22, 27, & 29 and put the sprout flow on tonight. #22 and 29, except beds 1 to 7, are planted with Honduras seed. #27 and beds 1, 2, 3, 4, 5, 6 of #29 are planted with Carolina seed, all at the rate of 2 bushels pr. acre. I ought to be hoeing, but I cannot get a sufficient number of hands to plant and hoe too, so I must defer the hoeing.

May 4. I commenced hoeing today in the settlement, but have very few hands.

May 9. The tides are very high. Champneys is flowed to within 2 inches of the stack gauge in the lower squares and some of the upper squares have all the water on them that they will hold. Mr. R. E. Briesenick of Brunswick stopped with us Thursday night. They were here in attendance on a court of arbitration in a case of Briesenick vs. the tug *J.C. Mallonee* of this place [Darien].

May 21. All my hoe hands "struck" on me today giving as a reason that "the work was bad, and they are afraid I will not pay them enough for it." I offered to pay whatever the work is worth, giving them my rule that a good woman ought to earn 6 1/4 cents pr. hour at hoeing, but they would not go to work. I then offered to hire them by the day until it could be determined what the work is worth, but no! So all left the field. I suppose some of them will be back to work tomorrow. The devil gets into them occasionally.

May 22. All my hoe hands who "struck" yesterday have returned to work as I expected, except two and I suspect one of these, Mary White, of being the source of the trouble.

May 26. I was re-elected County Commissioner for the 4th term and Academy Commissioner for the 4th term yesterday.

May 28. Mrs. James Baxter, who lived on Black Island near my home, died of erysipilas about 4 p.m. yesterday. Her funeral was held on Black Island at 8 o'clock this morning and her remains were taken by her husband to Jonesville for burial. Six negro men had to carry the coffin from the Black Island house to the hill at Ashantilly because the road was too bad to drive over. They are strangers here, belonging to the

Jonesville section of our county.

June 1. Isidore Collat of Savannah has re-opened [his] store in Darien today. He will keep shoes, clothing, hats and mens furnishing goods.

June 10. I borrowed Willie Strain's[176] naphtha launch and made a trip to Sapelo Light House yesterday evening to see a boat there offered for sale by Daniel Cromley.[177] I bought [it] for $20 and will convert it into a naphtha launch. It is about 26 [feet]. I will send for it tomorrow.

June 17. I am told that not more than 2/3 of a crop of rice has been planted on the Savannah and Ogeechee rivers because of the continued high water. I have not been able to plant #28 = 20 acres because of the continued freshets and Willie Strain [at Butler's Island][178] will have to curtail his planting for the same reason. C. O. S. Mallard and T. H. Gignilliat and D. S. Sinclair [all] up Cathead Creek have planted little or nothing and the negroes on Generals Island are not planting as much as they expected, so that the southern rice crop will probably be small this year.

June 19. Capt. James Dean,[179] a veteran Pilot of this port, died suddenly of apoplexy last night. He was buried today in the public cemetery on Cathead.

The Palmer Gasoline Engine ordered several weeks ago for a Launch for Champneys Island, arrived yesterday. It cost $108.60 delivered here. It is a 3 H.P.

June 26. I used the new gasoline launch for the first time this evening—she does well. The Cromleys named the boat "Shamrock" but I will call her the "Heron."

July 1. The new naphtha launch broke down yesterday and it took all day to repair it, so I did not get to the Island.

July 7. Claudie, Margie, Emma, Houstoun and Eloise Brewer will take a trip to the Seaboard tonight with the Walkers and Atwoods. They start from Capt. W. H. Atwood's place, "Cedar Point,"[180] this evening and will be gone all night.

July 29. I was confined to bed yesterday. This morning I found my launch sunk and have a great deal of trouble in getting it in working order again, although I had but little trouble in getting it up.

Aug. 1. I have just finished and put in an awning on our Gasoline launch. This makes the boat cost to date $217.00. I will have to have an Auto-Sparker. This will cost $15, so the Launch will cost in all, about $232.00. I have been using it new since June 19 [sic]. I have travelled about 150 miles and have used about 30 gallons of gasoline, or 1 2/3 pint of gasoline pr. mile traveled, including cost of taking care of the boat.

Aug. 6. The fact has leaked out today that Jos. Mansfield's daughter Eleanor—nicknamed "Pet"—is married, that she and lawyer Geo. Becket of Savannah were married last January and have kept it quiet even from her parents until now! This is the main topic of conversation in Darien today. It was not even suspected.

Aug. 24. I commence harvesting on the settlement today, but I have only 8 hands at this work, which is very discouraging.

Aug. 25. I have only 14 hands harvesting today! I do not know what is to become of the crop. I ought to have 50 hands, at least.

Sept. 19. Hands have not turned out and I am therefore suffering for the lack of labor. If I can get a [illegible] I think I can cut the loss down about $200.00. If not,

Railroad maintenance, Darien

the loss may be increased to an amount greater than $1200.00. Birds have been unusually bad but, as I have dried off all the squares to run them off, a great many—say 50%—have gone elsewhere. Then, one of our best mules cut herself on a wire fence and I will not be able to use her for some months. Truly, troubles multiply!

Oct. 24. Shipped today by Capt. Judkins about 2000 bushels of white rough rice. This is all I will have to ship! About 5000 bushels from 270 acres! This makes the yield from 270 acres threshed 19 bushels pr. acre when the scrap lots are counted. Statement. Shipped today 1892 bushels. Shipped previously 2989 bushels. Total 4881 bushels.

Nov. 12. I received orders from Mr. Cunningham to go ahead and prepare for another crop 2 days ago. They do not seem discouraged at the poor crop.

Nov. 14. Edward R. Poppell,[181] a well-known resident of the Barrington section of this county, died of Brights disease yesterday morning and was buried in the neighborhood in which he lived.

Nov. 30. I have sent Smith King to Bryan County to try to get some negroes who will live on the place [Champneys Island] and work for us regularly. I hope he will return tonight.[182]

Dec. 10. Hugh Thompson (Col) who has worked for me for 25 years died at 1 o'clock this morning. His death will prove a great loss to me.

Dec. 11. Ice again this morning—clear & still. The tide went so low this morning that I had to go to the Island via "3 Mile Cut," a distance of 12 miles, which I made in 2 hours.[183]

Dec. 31. December has been a remarkable month. We have had only two warm days since Nov. 25th last, and almost no rain. It has been a poor crop year in these parts. Cotton is selling in Savannah for 13 3/8 cents, middling. Rice is low, on account of the large Texas crop. Potatoes hard to get.

1904

March 2. I finish planting #1, 2, & 3 and put tonight's tide on them. #1 & 2 are sown with Gold seed raised on Rhetts Island (bought from W. H. Strain). #3 is sown

Ship loading lumber at Lower Bluff sawmill, Darien River.

with white Carolina seed of our own raising—all at the rate of 2 1/4 bushels pr. acre. My folks, including Claudie and the grandchildren, spent the day on the Island.

March 28. We have the intelligence this morning that Archie, son of my friend, Dr. Peter S. Clark of this place, died in Atlanta, Ga. last night of Grippe[184] complicated by malarial trouble. I suppose the immediate cause of his death to have been pneumonia. His father and mother were with him when he died. Poor parents! I feel for them! Archie was about 18 years old and was attending the Technological school.[185]

March 29. Archie Clarks remains were brought to Darien by the Str. *Hessie* today. His funeral was held at the Methodist church at the Ridge and he was buried immediately after in St. Andrews cemetery. I have never seen the community so wrought up because of a death in it. Our people all seem to feel a sense of personal bereavement. Dr. Clark tells me that he died of an old disease now called by a new-fangled name, "auto-toxhamine," caused by the "Grippe."

April 12. I took a trip on our Launch to Taylor & Bros. Lumber Mill in Glynn Co. about 25 miles from Darien. The following persons went with me. Willy Strain & Richard Strain, Rev. Mr. Smith and Houstoun. It was quite a pleasant trip. We went to see about buying some lumber. The "Heron"—the launch—made the trip of 50 miles in a little less than 7 hours, with the tide, but against the wind blowing about 8 miles an hour going. We used about 3 gallons of gasoline and about 1 pint lubricating oil so the trip cost about 75 cents.

April 21. Cloudy, with a stiff N.E. wind blowing and it feels cold enough to kill hogs. Temp. 57º at midday. Rice looks badly. The cold is too much for it.

April 26. My 52nd birthday. The wind blew hard from south yesterday. Today it blows about 20 miles an hour from the same direction and some clouds are gathering, so we hope it will rain. Vegetation cannot stand the drought much longer. The crop [is] suffering severely now.

May 6. The Board of City Commissioners elected me their clerk and treasurer yesterday. I therefore qualify this morning by giving bond for $2,000.00 with Willie H.

Strain and Jos. Mansfield as sureties and took the oath of office and took charge this morning. The plan pays $50.00 pr. month.

May 25. The dry weather continues—there are no signs of rain soon. All crops except rice are suffering for moisture. The drought is serious all over the south.

May 27. The Grand Jury have re-elected me County Commissioner and McIntosh Academy Commissioner.

May 31. Rained heavily several times during the evening, and we are thankful.

June 1. Birth—My 4th grandchild and 2nd grandson [Jesse Eugene Britt] was born this morning, son of Jesse and Claudie Britt.

June 3. The Episcopal Sunday School gave a picnic at Crescent today. They have wanted the other schools to join them. Margie and Houstoun go. Wife has been sick—mostly in bed for about six weeks and is still in bed with some rhumatic [sic] trouble.

June 8. I was re-elected Clerk and Treasurer for the City of Darien, unanimously by the City Board of Commissioners to serve for one year. I will now most probably hold the office as long as I want it.

July 28. I was appointed by the Ordinary one of five appraisers for the estate of Irvin Davis of this county. We left here yesterday, going to Jones Station[186] where we took dinner with Mr. C. O. Fulton and then took buggies and visited several lots of cattle and valued them, along with other articles belonging to the estate. We finished the appraisement Wednesday morning and made up our returns to the Ordinary and returned home last night. The value of the estate is $27,295.00, about.

Aug. 6. The rice crop is very handsome but I have so much volunteer in the best of it that I fear the yield will be greatly reduced.

Aug. 16. Last Saturday evening a flat loaded with beef cattle belonging to C. O. Fulton sank at the Darien wharf and 27 head of steers worth $17.00 each were drowned and Geo. Fuller fell overboard and was nearly drowned when pulled out.

Sept. 7. The wet, cloudy weather continues and rice is becoming over ripe for want of cutting, and it is falling in and down.

Sept. 21. I have found hands for the rice fields hard to get. So far fully 150 acres of the crop far over ripe before it could, or can, be cut. I now have 115 acres of uncut rice—all ripe. I estimate our loss from over ripe rice at 1000 bushels. Fortunately the rice birds have not been as plentiful, nor troublesome as usual. I have threshed about 500 bushels of rice so far.

A few days ago, Mr. S. H. Synot, who sells the Champneys Island rice, wrote that he had sold all of our rice on hand at 62 cents pr. bushel. He wrote me early in the season that ours was the best rice received in the Savannah market.

Oct. 4. I made my first shipment of rice this evening—about 2389 bushels by Capt. Judkins. It is white rice from #3, 5, & 7, about 70 acres gross—65 acres net.

Oct. 19. I have a Lighter loaded with rice at Champneys Island that has been ready for shipment ever since the 14th inst., but Capt. Judkins has not yet come for it. I suppose he is detained on the Satilla by the stormy N.E. winds that have blown for the last 7 days.

Nov. 14. Rain lasted until around midnight when the wind shifted to west and blew probably 40 odd miles an hour. This morning temperature was 38º so we had quite a heavy frost, being the first worth mentioning for the season. It is clear & still and so by no means unpleasant. The tides are so low that I could not get to the Island this morning.

Dec. 13. I have advices from Mr. Cunningham that the Central Ry. people will probably not plant Champneys Island next year, but will offer it for sale. This is not as yet certain.

Dec. 20. Some one broke into the mill building on Champneys Island last night and stole about 10 bushels of rice. They also stole 6 or 7 of my fowls. I therefore put on the night watchman again.

1905

Jan. 3. Champneys Island was sold in Darien by the Central of Georgia Railroad Co. to William C. Wylly for $2,000.00, about 1/3 of its value! And so I am, or will soon be, out of that job. I have the personal property in hand, yet to be sold. Mr. Cunningham and Col. A. R. Lawton came down to sell Champneys Island. [Editor's Note: McIntosh County deed records show that William Strain was the actual purchaser of the island, although William Wylly may have been acting as Strain's agent in this instance. See the Introduction, note 17].

Jan. 10. Mrs. Shine[187]—an old inhabitant of Darien and an old lady, 83 years I think, died and her remains were brought to Darien today and buried at St. Andrews Cemetery.

Jan. 31. I shipped all the rice on Champneys Island to Savannah by Capt. J. H. Judkins. I also shipped 690 bales straw to Mr. C. N. Roberts, Savannah.

Feby. 13. W. C. Wylly had his collar bone broken in a run-away scrape yesterday in which the buggy in which he was driving was upset on the Shell road.

March 27. R. Austin Young's[188] dwelling at Kell's Grove was destroyed by fire Saturday night. He lost all his furniture & effects.

April 3. The community is greatly shocked by two distressing occurrances—1st by the drowning of Hyman Bluestein, as told in the newspaper clipping pasted here.[189] 2nd, by the following: A Mr. Copeland, from somewhere near Liberty County, moved his family, consisting of his wife and 5 children, the eldest being 8 years old, some weeks ago to the Barclay place on the Ridge. Copeland works at the Canning Factory at Shell Bluff.[190] He left here last Monday morning to be gone a week. He returned home last Saturday night. On his arrival, he found Mrs. Copeland dead and the poor little children all asleep. On awaking them, the eldest, a little girl of 8 years, told him the following pitiful story—that her mother had taken a dose of calomel[191] Friday night and had awaked her about 2 o'clock Saturday morning saying that her heart hurt her and she was dying, and that the mother then died and that she and her little brothers and sisters had remained with the dead mother all Saturday without anything to eat! When asked why she had not notified some of the neighbors she said she was afraid to leave the little baby and could not send the next eldest, her brother, because she was afraid to stay by herself. The nearest neighbor is Mr. McD. Dunwody, about 200 yards in a direct line, but about a fourth of a mile by road. What the poor little things suffered can better be imagined than described. Mrs. Copeland's youngest child was a baby at the breast, and it is thought sucked the bosom of the dead mother.

Shellbluff oyster cannery at Valona, 1906.

The whole story is so sad that it is hard to write it. Mrs. Copeland's remains were buried at the Public Cemetery yesterday.

April 26. Sun Brothers Circus is here showing today. They gave me several complimentary tickets, but my headache prevented my attending. My 53rd birthday! I am celebrating it with a headache, without a "jag" on. All of the children attended the "Show" at night and all were delighted. I have never seen a circus and wanted to see this one very much.

The Darien & Western Railroad ran their first passenger train to and from Ludowici (Lot-o-Whis-Key!) on the Atlantic Coast Line Ry. on last Monday and they are making trips to that point regularly now, every day. They have also extended their line to Lower Bluff Mill where they have constructed a wharf.[192] They expect to extend their line to a point on Black Island, on "Long reach," Darien River.[193]

May 21. Another day of bad news and disasters. George Fulton was drowned near Lower Bluff Mill Saturday evening and after dragging for his body all night it was recovered about daylight this morning. He was subject to epileptic fits and it is supposed that one seized him and he fell off a flat that he was building a cattle guard around and was drowned before he was missed. His remains were buried in St. Andrews Cemetery. He was the son of Mr. C. O. Fulton.

Wife has been quite sick—confined to bed for several days and is still in bed and under the care of Dr. Clark, but is better this afternoon.

June 20. I was elected Clerk of the Board of County Commissioners today by a mere majority of the Board. I have therefore resigned my commission as a member of the Board of Commissioners for McIntosh County and I will take charge of the Clerk's office on the 1st of July.

Aug. 24. I bought today from Robert A. Strain[194] the gasoline launch "Heron" that I had fitted up for Champneys Island in June 1903. I paid $85.00 for it and will put it in good condition at once.

Nov. 10. Rain has been falling continuously since midnight last night and indications are that it will continue all day. This is the first all day rain in two years and we hope it extends to points up the river so that some timber and lumber can be floated down this way. This community, in common with Brunswick & Savannah, has lately been sorely afflicted with "Dengue" fever. There have been many cases—the number running into the thousands in Brunswick. Probably one half of our white people have had it, or have it now.

Nov. 17. The war between Japan and Russia is over, Japan having gotten by far the best of it, having captured or destroyed Russia's navy and having captured all of Russia's possessions in Manchuria and half the island of Sakhalin. But Russia's troubles are not yet ended, nor is the end in sight. There are riots, strikes and attempted revolutions there almost every day and the world is wondering what will happen next.

Mrs. Reuben K. Walker died last night at her home at Crescent after a long illness. Her remains will be buried today in the Hopkins burying ground on "Belleville" plantation.

Dec. 5. Mrs. Caroline Donnelly, probably the oldest citizen of Darien, aged 83 years, died at her home in Darien of fever this afternoon.

Dec. 19. Mr. Thomas H. Gignilliat died this morning at his house 5 miles from Darien of some heart trouble. He has been ill for some time. The funeral will take place tomorrow from the Presbyterian Church and his remains will be buried in St. Andrews Cemetery.

1906

Jan. 22. A dreadful tragedy was enacted in Darien this evening. The Russian Barkentine *Capella* is here loading with lumber. A boat and three sailors were sent to Darien for supplies. While waiting at the wharf near Mansfield's wharf, a drunken up-country man named Clements went down there and, without any provocation, shot one of the sailors dead, then shot another through his hat. This man jumped on board and got behind the boat when Clements shot him through his hand

Clements then went up on the main street where he met W. W. Sinclair, the marshal, and shot at him twice. Sinclair returned the fire, shooting Clements in the lower bowels. We cannot but hope that the marshal's shot will prove fatal because Clements having murdered an inoffensive obscure sailor and there being absolutely nothing to justify the act—a jury will probably acquit him on account of the prevailing maudlin sentiment universal now. It is almost impossible to convict a person—unless they are innocent. The marshal was unhurt but had a narrow escape.

Jan. 24. By Mrs. Strains permission I had some cattle put on Champneys Island last Saturday. The man—Dan Grant—I employed to put them over left my cow "Hilda" in the bog. She was later taken out Sunday morning by Mr. Dolbow's[195] fishermen, but I fear she will not live. I think of prosecuting Grant for "cruelty to animals."

It is thought that "Bud" Clements, who killed the Russian sailor Monday, will recover.

Feby. 20. Mr. J. Alfred Atwood[196] died at Shell Bluff of paralysis on last Saturday night and was buried from the Presbyterian Church in the "Upper Mill Cemetery"—recently owned by the Presbyterian Church.

April 25. A most dreadful disaster befell San Francisco, Cal. on the 18th inst.—the final particulars have not yet reached us, but the following is known. A violent earthquake destroyed or damaged a large number of buildings and the [illegible] took

Hilton-Dodge Lumber Company sawmill at Lower Bluff, ca. 1906.

fire. From 2 thirds to 3 fourths of the city is burnt and about 200,000 people are rendered homeless. The estimates of the death rate vary greatly. The U.S. authorities estimate the dead at 277, while the coroner estimates the dead at 1000.

Oct. 20.   We are having a stormy time of it. There have been several cyclones this week. On Thursday, great damage was done to life & property in Florida and Cuba. A N.E. gale has been blowing here this week and tides have been higher than since Oct. 2nd, 1898 and [have done] considerable damage—Rhetts & Butlers islands each have several bad breaks and nearly all of the islands in this delta were flowed more or less, and the negroes who planted Rhetts Island lost about all of their rice.

Dec. 19.   The second long disastrous drought of 1906, lasting from the middle of October to yesterday, seems to be at an end. A light, drizzling rain has been falling since yesterday morning.

Dec. 24.   Like last December, this month has tried to give us a little of all kinds of weather. The weather is so changeable that colds are prevalent (and so is mumps) and we feel the sudden changes keenly.

Dec. 28.   I have been barely able to go for 3 days from a severe cold. Wife is also very unwell and the grand children are all more or less sick from colds.

1907

Jan. 3.   Mr. William A. Wilcox[197] died quite suddenly late yesterday evening of pneumonia. He was an ex-state senator and quite a prominent man, aged about 70 years.

Jan. 24.   At the last regular meeting of the Commissioners for the City of Darien I was elected a committee of one to visit and report on the systems of street paving, water works, electric lights and sewerage at Douglas, Ocilla, Abbeville, Helena, McRae and Baxley.

I therefore left here on my itinerary on the Str. *Hessie* on Monday afternoon. I stayed one night in Brunswick, stopping at the Oglethorpe Hotel. I left Brunswick Tuesday morning by the Atlanta, Birmingham and Atlantic Ry. (which I believe is the first road in the South) and reached Douglas about 11 a.m. Douglas is a fine, progressive town of about 2000 inhabitants. After a thorough examination of the public utilities there, I left Douglas for Ocilla that evening, arriving at Ocilla about 9 p.m.

Wednesday morning I had a blooming headache, but managed to keep going. Ocilla is also a fine growing town—having about 2500 inhabitants. At both Douglas & Ocilla I found the authorities very kind and accommodating. Having finished my investigation there, I left for Abbeville, arriving there about midday. Abbeville is also a growing town, rather older than the other places visited. I left Abbeville Wednesday evening for Savannah, arriving there that night. I did not think it necessary to stop at Helena and McRae, nor did I visit Baxley. I think the information gathered at Douglas and Ocilla all that will be needed by the Darien folks. I left Savannah Thursday (today) morning, arriving home about 11 a.m. after a trip that I greatly enjoyed even in spite of my headache on Wednesday and bad feelings today.

March 29. Hon. F. H. McFarland, Chairman, Board of County Commissioners, appointed me representative of the McIntosh County Board of Commissioners at the convention held in Savannah yesterday in the interest of a good road from Savannah to Darien. I attended the meeting at which were representatives from Chatham, Bryan, Liberty and McIntosh counties. It was determined that Chatham County is to extend and surface the Ogeechee Road to King's Ferry and they are to bridge the Ogeechee River. Bryan County is to build and surface from the causeway approaching the bridge in Bryan Co. and surface the Stage Road to the Liberty Co. line. Liberty Co. is to grade and surface the Stage Road from the Bryan Co. line to the McIntosh Co. line on South Newport [River] bridge. And McIntosh Co. is expected to grade and surface the Stage Road from South Newport bridge to Darien [18 miles].

April 9. Jesse Britt took little Legare to Savannah yesterday to have growths removed and took Charlotte along in order to have her throat and ears examined because she seems to be getting somewhat deaf. I got a telegram from him this morning saying that Dr. Julian Chisholm had charged $100.00 for the operation which seems to me to be highway robbery.

April 11. J. A. Britt and children have not yet returned home, but we hope they will come in tonight. Both Charlotte and Legare were operated on for the same disease—adenoidal growths in the nose or throats and were doing well at the last report.

I should have recorded the fact earlier that the stock holders of the Darien Ice Mfg. Co. elected me a Director at their meeting Feby. 26th last and on the same day the Directors of the Company elected me secretary & treasurer. The position I believe pays $130.00 per annum.

April 19. Our pastor, the Rev. N. Keff Smith, D.D. having tendered his resignation to Presbytery without good reasons—I was sent by the Darien church to Presbytery of Savannah which convened in Blackshear on the 16th inst. I attended the opening of Presbytery that night and the session next morning at which I opposed the resignation on behalf of the church. Presbytery refused to accept his resignation and so Mr. Smith will remain with us, but under a different arrangement. Heretofore he has given us 3 Sundays each month. Under the new arrangement, he will give us 2 Sundays each month and he will serve the Dorchester & St. Marys churches the balance of the time.

May 14. Last Saturday a negro man—unintentionally I am told—shot and killed a young white man named Sands at Darien Junction. The negro was caught and put in

Georgia Coast & Piedmont Railroad shops and depot at Crescent, ca. 1907.

jail here. Last night a rumor reached the authorities here that a large number of men from Glennville—the young man's home—thought of making an effort to lynch the negro, and the authorities ordered all colored prisoners taken to Brunswick last night for safe keeping. Today, about 50 men came in fully armed on the passenger train, which they seized for the purpose. They got here about 2.15 o'clock, and went immediately to the County Jail, which they found open for their benefit. They returned by the passenger train, which they held until they were ready to go. They left here at 2.45 p.m.

The people are very much wrought up over the outrage and the Governor has been informed of the matter by telegraph. I am writing this immediately after the departure of the train and am ignorant as to whether they did any harm while here. (They fired a great many shots as they were going out).

Aug. 6. The Co. Commissioners very kindly raised my salary from $40.00 to $50.00 per month today.

Aug. 31. Our church has suffered another loss in the permanent removal from this community of Elder C. Reid Walker, who left for Waycross this week.

Sept. 5. C. R. Walker, who in the last note is said to have left Darien permanently, returned on Monday and intends to remain. He did not like Waycross—nor the striking telegraph operators there.

Oct. 7. Mr. Arthur Bailey's[198] store on the Ridge was destroyed by fire Saturday night. Fire accidental—loss quite heavy, no insurance.

Nov. 11. Col. John H. Estill, owner of the Savannah Morning News, is dead! He died Saturday night last and I feel a sense of personal bereavement—having read his paper and editorials for 30 years and having met him personally.

Dec. 23. We have been having a hard time of it at my house, with La Grippe. I was taken sick on the 7th inst. and am not well yet. Then wife, Houstoun, Emma, Margie, and Claudie were taken with it. For several days there was no one well enough to cook for the rest. But we are on the mend now.

1908

Jan. 1. Emma and Dan L. Whitesides[199] were married at the Presbyterian Church today and left on the steamer *Hessie* for their bridal trip to Florida points.

Jan. 30. Mr. John Muller, one of our older citizens, died in Savannah last night after several years illness—paralysis I think—his remains will be buried near his wife on Harris Neck in this county, tomorrow.

Feby. 4. Dan Whitesides & Emma have gone to house keeping in the Ben Sinclair house in Darien. They moved in today. I expect them to be in the house next door to ours in less than a year.

Feby. 12. Harvey T. Long [Editor's note: see the next two entries], one of my oldest & best friends, died yesterday evening of Grippe complicated by pneumonia. His remains were taken to St. Marys, Ga. for burial. My old friends are rapidly passing over the river!

Feby. 13. The store of the R. A. Strain Co. is open again but Harvey Long is not there! I—and others too—will greatly miss him. I feel a sense of personal bereavement. He was one of the comparatively few who "do things" and do not merely talk. He was, at the time of his death, Chairman of the Board of Education and an Academy Commissioner and a Pilot Commissioner for the City of Darien. A few years ago he was a City Commissioner and one of the best the City ever had.

Feby. 27. Willie Strain has returned here to live. He will take Harvey Long's place as manager of the store of R. A. Strain Co. I am glad he is back again.

Feby. 28. Hanging! Lee Holmes, colored, who killed Dr. Sands at Darien Junction May 11th, 1907, was hanged in the jail yard at mid-day today, the execution having been in private. His people, who live at or near Harris Neck, took his body home with them for burial.

Aug. 25. I killed a Rattle snake about 4 ft., 8 inches long, having 12 rattles near the culvert today, being the largest I ever killed and as large a one as I ever saw. I will keep his rattles.

Sept. 8. Dr. Spalding Kenan died at his home in Darien on Saturday evening of diabetes and acute indigestion. He was 73 years old. His remains were buried in St. Andrews Cemetery yesterday morning.

Sept. 26. Mr. T. Butler Blount—ex-Sheriff—died at his house on the Ridge at 2.30 o'clock this afternoon of cancer of the stomach. He was 66 years old.

Oct. 30. The Rev. Nicholas Keff Smith, D.D. has resigned the pastorate of the Presbyterian Church in Darien, to take effect tomorrow. He will serve the James Island Church in Charleston Presbytery. I greatly regret that circumstances are such that he is compelled to leave Darien, and us.[200]

Nov. 17. Law suit. J. Bluestein, Agt. vs. J. G. Legare—tried today before E. G. Cain, J.P. Cause of action: About 3 weeks ago I employed Bluestein to haul off a dead cow from my yard, promising to pay him $2.00 for hauling the cow a sufficient distance across the Shell road so that the smell should not be offensive. After hauling the cow out, Mr. Bluestein skinned the cow against my wishes and sold the skin. I claimed that the skin was sufficient compensation and plead Sec. 226 of the Criminal Code of 1895, which states that it shall be unlawful to skin an animal belonging to another without the consent of the owner, &c.

I lost the case—Justice Cain deciding against me on the following points: 1st, one cannot plead or quote a criminal law to prove ownership in a civil suit! 2nd, I relinquished all claim to ownership when I delivered the dead cow to Bluestein. 3rd, He sustained the contention of the plaintiff's attorney, W. A. Way (a lawyer to whom Judge Seabrook administered the following dreadful rebuke in the open Superior Court—"Mr. Way, I wish you would be perfectly frank with me") that Sec. 226 was enacted for the sole benefit of railroads, which of course would have made the law unconstitutional. Mr. Cain would not allow me to put a witness on the stand! These things I know to be unlawful, unjust and oppressive, but I will submit rather than go further *in his court.* But, "Every dog has his day," and "the cats have 2 afternoons." Things will soon even up and right themselves after a while.

Dec. 9. A son [Joseph Legare Whitesides] was born to D. L. and Emma Whitesides today. This is my 5th grand child and 3rd grandson.

Dec. 30. Mr. William J. Britt, father of Jesse Britt, died at Crescent last night and will be buried this afternoon.

1909

Jan. 19. Lee's birthday—holiday—We had a family dinner at home, celebrating wife's and Charlotte's birthday (Jan. 18th). All members of my family were present. We also had our new minister, Mr. & Mrs. W. S. Milne,[201] with us.

Legare family portrait, ca. 1904. The girls in front (l-r) are Charlotte and Emily Britt. Seated (l-r), Margie Legare, John G. Legare holding John Legare Britt, Charlotte Legare, Claudia Legare Britt. Standing (l-r), Houstoun Legare, Emma Legare, Jesse A. Britt.

Jan. 26. Governor Joseph M. Brown is here on a visit. He arrived on the steamer *Hessie* yesterday midday and is the guest of the city. He was entertained with an oyster roast & fish fry at the Court House last night. Today he is being taken over the harbor in a steam boat, accompanied by many Darienites.

This entertainment was so strangely managed that, although I am told that the City of Darien will bear the expenses of it, yet I—as Clerk—know almost nothing of it and cannot find out. I was not invited or requested to attend. I am told that it was a public entertainment but I was ordered to invite, and did invite, officially, the county commissioners to participate. I suppose I will hear all about it next week when the two boards meet.

March 22. My little grandson, Joseph Legare Whitesides, was baptised by the Rev. Wm. S. Milne at Sunday School at the church on yesterday afternoon. The little fellow is growing finely and we think him unusually intelligent.

April 2. Mrs. Augusta J. Pease,[202] the oldest member of our church, died at Palatka, Fla. of acute indigestion, followed by paralysis, yesterday morning. Her remains were buried here in St. Andrews Cemetery today. She was 81 years old.

June 17. The First Presbyterian Church of Darien celebrated its centennial anniversary this week in the following manner—

Sunday. Rev. Wm. S. Milne preached morning, afternoon and night.

Monday. Mr. Milne preached at night.

Tuesday night. Rev. N. K. Smith, D.D.

Wednesday night. Rev. J. G. Fair, D.D. preached a fine sermon from Exodus 3– Vs. 2, "And behold the bush burned with fire and the bush was not consumed."

The congregations were unusually large and there was considerable interest manifested and we hope that good was done.

Oct. 21. In coming from Crescent last night, a log train lost several cars. Mr. J. L. Hines,[203] the Superintendent of the G.C. & P. Ry. was following the train on a motor car which ran into the wild car, injuring Mr. Hines dangerously and breaking the leg of Mr. Owens who was with him. It is feared that Mr. Hines will lose one leg. His forehead and mouth were broken in. Mr. Hines was taken at once to Savannah for treatment.

3 p.m.—We have just heard that Mr. Hines is dying from his injuries.

Mr. Hines died at 10 p.m. His remains will be buried at Hinesville, Ga.

Nov. 10. We have just heard of the sudden, and unexpected, death in Savannah last night of Mr. O. C. Hopkins, familiarly known as "Bully" Hopkins. Mrs. Wm. Henry Atwood died at her home near Crescent on the 1st instant.

Nov. 13. I had my photograph taken in Darien today, by request of my children and grand children.

1910

Feby. 12. Fire in Town—Clarke Brothers offices, 3 buildings on Broad St., were burnt about midnight last night, cause of fire unknown. One of these offices was occupied as the U.S. Custom House. The government property was saved; the Clarkes lost everything in the building. But for the fact that a rain had fallen during yesterday, this would have been a far more serious fire to the town.

Feby. 17. Fire—The Steamer *J.C. Mallonee* was burnt at her dock in Darien last night, the fire being discovered about 11 o'clock, cause of fire unknown. Another great loss to the town. As yet, I am uninformed as to whether it will be replaced by another boat. I hope it will be.

March 10. My sixth grand child—a daughter—[Claudia Whitesides] was born to Emma and Dan L. Whitesides, at Ashantilly, this morning.

April 4. This is quite a notable day because of the fact that 22 Automobiles visited and passed through Darien in an endurance run from Savannah to Jacksonville, Fla. The Automobiles were entertained by the citizens of Darien.

April 26. My 58th birthday. Everything eatable is very high in price. Meats are higher than I have known of their being since 1865.

May 7. On Thursday afternoon a S.A.L. [Seaboard Air Line] train ran over Arthur W. Young near his house at Jones and injured him badly. He was drunk and lying to one side of the track with his head on a cross tie and when the train passed over him it missed him and he raised up—and the ends of the axles to the cars struck him. He is in a hospital in Savannah and I am told that his chances for recovery are good, but he will be badly disfigured and maimed. We hope it will result in his giving up liquor drinking.

June 3. Holiday—I took the grand children to Doboy Island today for a little recreation. We had a pleasant day of it.

June 21. Mr. F. G. L. Grundy and Miss Louise C. Hopkins[204] were married by Rev. Samuel French at the Episcopal Church this evening.

June 27. An old friend and neighbor, Mr. Henry A. Weil, died at his home in Savannah last night, quite unexpectedly.

June 28. The City and County authorities, having granted me permission to go away for a few weeks for my health, I leave Darien today somewhat uncertain as to my destination, but I am going to the mountains of North Carolina. I leave Houstoun in charge of my affairs.

I arrived at Columbia this evening and was met at the depot by cousin Ned Girardeau who took me to his home to spend the night.

July 12. I am still in Columbia. Ned G. took me over the city. I left Columbia at 1 p.m. and arrived at Ninety-Six, S.C. at 4 p.m. and was met there by Geo. Holland and Miss Mary, daughter of Henry Holland. I called on old friends of the family. They all treated me as kindly as it was possible for them to do, which made me feel very much at home...

...I like Waynesville [N.C.] so well that I have concluded to spend my vacation here. I saw several real live Indians here today—they belong to the Cherokee tribe whose reservation is only a short distance to the southwest of Waynesville.

July 21. I visited "Eagles Nest" Hotel and the north view on the Junaluska Mountain today. Eagles Nest Hotel is 5050 ft. above tide level and about 2300 feet above Waynesville. To reach it one has to travel a well-graded beautiful mountain road that ascends fully 2000 feet in about 3 miles. The scenery is grand, and will amply repay anyone for the trouble and expense incurred in making the trip.

July 24. I attended the Methodist church today and noted a most distressing thing—*the Sunday School children going away from the church service.* In that country they have Sunday School in the morning, before the morning service and when the school turned out the Sunday School scholars left and went home, regardless of size! I fear that this is the practice in the other churches [here] as well as the Methodist church! Alas for the community! The sermon and the singing were good but this congregation aped the Episcopalians worse than the Presbyterians.

July 28. I left Waynesville on my return homeward at mid-day, and having to lay over about 3 hours in Asheville, I spent the most of my wait in "doing" Asheville.

July 29. I arrived in Darien at 8.15 a.m., having travelled all night, and I feel like a "biled owl." I find that my business has been well cared for and all are well. I am glad to get home again.

Aug. 8. I have not had a severe headache since I left home July 11th last and, as I sleep well and eat heartily, and feel well, I conclude that my trip did me much good.

Aug. 25. We are in the midst of a very rainy spell—I think that fully 8 inches of rain must have fallen since last Saturday. It has now been 7 years this month since the series of dry years set in and I am anxious to note whether the weather does or does not work in cycles of 7 years duration. My impression is that it does. (Not proven in this case).

We have just passed through a most exciting political campaign for the Governorship of Georgia, the candidates being Gov. Jos. M. Brown and ex-Gov. Hoke Smith. Mr. Smith has just been nominated by a small majority—said to be less than 5000 [votes] in the entire state.

Aug. 29. A very peculiar thing has happened here. About 20 years ago the white people of this community bought a Hearse for the use of the white people of the community, and the City of Darien built a house on their own land to house it in. Now, about a week ago, *one* of the hearse trustees (I am a trustee for the hearse, but know nothing of the move) concluded that the hearse-house was not just what he wanted. So he told the City Marshal that if he—the Marshall—would take the building down, he could have the lumber for his trouble and expense. The Marshal proceeds without further authority and took down the City's property—the hearse house— and hauls it away. The Mayor and Councilmen each saw what was going on, but thought that some of their number had ordered it done (illegally?) but they did not care to interfere! Now the question is—what are they going to do about it? And, could such a thing have occurred anywhere else than in *poor old, dead Darien!* [Editor's italics].

Sept. 5. Frank E. Durant,[205] ex-County Treasurer, died at Meridian this morning—Brights disease—and will be buried in St. Andrews Cemetery at midday tomorrow.

Sept. 13. Mr. William S. Mallard[206] died last night at his home in Darien of Locomotor Ataxia, which he had been afflicted with for many years. His remains were buried in St. Andrews Cemetery this evening, the funeral having been preached in the Presbyterian Church.

Sept. 16. I see by the [Savannah] "Morning News" that N. Heyward Barnwell died in Brunswick yesterday and was buried in the cemetery there. He was an old friend—but a very peculiar man. He is responsible for my coming to Georgia 33 years ago—I came to work for him. He died of malarial trouble, aged 70 years.

Sept. 24.   Capt John Hagan[207] died at his home on the Ridge of pelagra this morning, aged about 65 years.

Oct. 19.   **Storm—**After several days warning a severe N.E. gale struck us yesterday. The wind had been blowing quite hard for several days and we had an occasional rain. 10.30 p.m., Barometer 29.25, wind blowing in gusts of about 60 miles. Tide at Ashantilly about 2 ft [above normal] and at Darien about one foot higher than I have ever seen it except on Oct. 2nd, 1898.

Results—The telegraph lines down to the south of us. In Darien there are a number of trees blown down, or broken, otherwise I have not been advised of any great injury. I had my fences slightly damaged and some shade trees broken, but not badly and lost some green oranges. The rice places have suffered severely. I cannot at present even guess at the extent of the loss.

The steamer *Hessie* from Brunswick has not been in two days so there must be big trouble in Brunswick.

Dec. 9.   My dear old friend, Capt. James Houstoun Johnston died very unexpectedly on yesterday morning of heart failure, superinduced by a cold, in the 80th year of his age at his home in Savannah. He was like a father to me!

Dec. 10.   Samuel Randolph Dean,[208] Vice-Mayor of Darien and an old and valued friend of mine, died at his home in Darien this morning from Brights disease after an illness of 2 weeks.

Dec. 16.   I have never known of so much illness in this community among the white people as there has been during the last 60 days. D. S. Sinclair had a stroke of paralysis 3 days ago and is at the point of death.

1911

Jan. 11.   Mrs. W. J. Britt died in Waycross, Ga. on Sunday morning last and was buried there. (Pneumonia).

Mr. S. W. Williams,[209] who lived at South Newport, an ex-County Commissioner, died in Savannah of brights disease on yesterday and was buried near his home today.

Jan. 17.   A thing before unheard of here has happened. John W. Williams, who worked for the Darien Ice Manufacturing Co. under my direction—in 1909, and last year—conducted a livery stable in co-partnership with Mr. John P. Mew[210]—doing business as Williams and Mew—forged a draft for a small amount and abandoned his wife and little baby, and has now been gone a week and apparently no one knows where he has gone to or what has become of him. Mr. Mew got injured on the G.C. & P. Ry. about 3 weeks ago and is now ill in a hospital in Savannah.

Williams left 6 horses and mules in their stable that were without feed for 3 days, after which the City has been feeding them. He left his wife and baby in destitute circumstances and friends are now caring for them.

Feb. 15.   The Darien Telephone Co. has commenced to erect poles in the streets today. I hope it means that we will soon have a telephone system over the town and community.

March 7.   Capt. A. C. Wylly died today of general break-down from old age, aged 78 years. He was one of that fast disappearing type of old-time southern gentlemen. His funeral will take place from the Episcopal Church and his remains will be buried in St. Andrews Cemetery immediately after.

March 9. I am covering the L part of my house with roofing felt–hope to finish tomorrow. Roofing with felt is something of an experiment but I hope for good results. I cannot obtain really good shingles now at a reasonable price.

April 1. Two deaths in the community! Col. Wm. Clifton[211] died at his house on the Ridge of nephritis this evening, aged 56 years. Dan Fynn died of pneumonia earlier in the day. He was about 22 years old.

April 15. Mrs. J. S. Fynn died at her home on the Ridge today of nervous breakdown from nursing her 2 sick sons, one of whom died two weeks ago and one is recovering.

April 17. Mr. J. S. Fynn died of pneumonia today after a two weeks illness–making the 3rd death in this family in two weeks and we are informed that two other members of the family are now ill.

My old friend Joseph A. Walker died in a hospital in Savannah today.

April 26. My 59th birthday! I and my children and grand children, except Houstoun, took a trip around Generals Island and took dinner at Butlers Island, getting home about 5 o'clock p.m.

May 26. C. O. Fulton, Sr. died very suddenly, away from home, at midday today of some heart trouble. It is a great shock to us as he was in town and with me on the Grand Jury on Tuesday. He was driving cattle and stopped to rest at the home of Mr. Dave McIntosh who lives about 3 miles from Mr. Fulton's house at Jones–and laid down to take a nap and died in a few minutes...Alas! My old friends are rapidly passing over to the other side of the great "divide."

I was on the Grand Jury this week. They–the Grand Jury– elected me Justice of the Peace for the 271st District, with Mr. E. G. Cain resigned. I fear the office will be troublesome and pay little.

May 29. I have bought for wife 40 shares of the stock of the Darien Ice Manufacturing Co. from J. A. Walker's Estate and I have bargained to take the balance of his holding, 75 shares, to be paid for at $5.00 pr. share as soon as I am able to pay for it, and it is to be delivered to me as paid for. Until paid for the 75 shares are to be held by R. K. Hopkins and he is to act as one of the Directors and he will vote with me on all reasonable propositions–thus giving me control of the affairs of the Company. I am authorized to borrow $300.00 to be used in repairing the plant, which has now been idle for 18 months. I therefore take charge and go to work at once to get the factory in running order. I find the machinery in bad condition but hope to repair it so as that it will run this season.

June 16. I have had a great many back sets [setbacks] in repairing the Ice Factory, but hope to get to work today sometime. I have hired a colored man named Albert Houston to take charge of the machinery and run the plant and Houstoun Legare will sell the ice and collect the money. I am hoping and praying that the venture will prove a success.

June 19. The Factory is still working, but we have no ice yet. Last Saturday the water supply gave out, causing us to lose 7 hours–then the ammonia pump stopped on us at midnight Saturday night and we did not get it to work again until 1 p.m. Sunday. (I very much regretted having to work Sunday, but it could not be helped without great loss). The machinery has been working very well since.

June 21. The Ice Factory is giving so much trouble from numerous small defects that I have stopped it and am making necessary repairs, which I hope to finish tomorrow.

June 28. Thomas A. Bailey, Mayor of Darien, and Miss Sarah Scott of the Ridge were married in Savannah today and left for parts unknown. The whole proceeding was done so much on the quiet that I almost feel guilty in recording it. Just when they will return, no one knows, but I have an idea that they are going to catch it when they get back. I think the secrecy observed was to avoid an "ovation" from their friends but my idea is that the "ovation" will be given all the same. They are both very much liked in the community.

July 25. I was compelled to shut down the Ice Factory to repair some bad work.

Aug. 15. After a constant fight to keep the old thing going, I shut the Ice Factory down for an indefinite period on Friday the 11th instant. Wife & I have apparently lost heavily by our venture in trying to buy and operate the Factory. I do not know what will become of it, time alone will tell this.

Aug. 22. A meeting of the stock holders and also of the Directors of the Ice Factory was held today at which the Factory was ordered sold to pay its debts on Sept. 5th next.

Oct. 31. Rainy, gloomy all day and there is more bad news afloat than I have heard in a long time—1st, we hear that Mr. D. M. McIntosh lost two sons by death last night. We have no particulars except that they had bilious fever last week.

2nd, August Schmidt[212] & Co.—we hear—are about to fail and lose everything.

3rd, That the R. A. Strain Co. is tottering! And still it rains.

Nov. 27. Yesterday, while ferrying Automobiles across the [Altamaha] river from Dents to Darien Alex Bailey overloaded the lighter, pulling 4 large Automobiles on it and the lighter sunk in the mouth of Generals Cut, one large new machine falling off in the hole there—which is said to be 30 odd feet deep! And all of the machines were covered with water—and, of course, badly damaged and, I am told, he jeopardized the lives of his passengers among whom were several ladies, by telling them there was no danger! And that, but for the timely passing of another launch & lighter belonging to a Glynn Co. gentleman, which stopped and took the passengers off, some lives might have been lost. I hope all of this is not true, but fear that it is.

I hear that the wrecker "H. Toomey" now here, has been employed to rescue the Automobiles. I cannot but dread the outcome of this accident. Our part of the country has a name bad enough without such an unnecessary accident to help blacken it. My impression is that this will badly injure our chances of locating the Florida & Northern Automobile highway on this route—which we are trying to do.

1912

Jan. 22. The Sea-going Railroad—the Florida East Coast Extension to Key West— is opened for trafic [sic] today. It is considered one of the most wonderful engineering

Walton Street, looking south toward waterfront, 1912. G.C. & F. rails are being laid.

feats ever attempted. It is built of concrete mostly and extends 78 miles across water, most of it open sea.

March 22. I made a trip to Savannah for the County to buy them 4 mules, 2 wagons, one cart and necessary gear. I went up on yesterday, taking Eugene Britt, my little grandson, with me. I had the only bad headache I have had in about a year today.

March 30. I am building a rowboat for use in fishing &c. on the salt water at home.

Legare Britt is down with a case of measles.

April 18. Last Sunday night the Steamship *Titanic* the largest ever built, on her first voyage over the sea, from Liverpool to New York, struck an iceberg and sunk, drowning 1635 of her crew and passengers, among them quite a large number of prominent and very rich people. This is the worst marine disaster on record. The survivors have not yet reached shore.

April 22. Rained off and on all yesterday and last night. Today is cloudy, but up to midday no rain has fallen. 4 8/10 inches of rain were recorded Friday evening until Sunday evening by the U.S. Weather observer Mr. J. M. Atwood[213] at Valona. I don't ever recollect seeing so much rain, thunder and lightning in the spring of the year.

Details of the *Titanic* wreck are now coming in, but the death list is not diminished. While the whole country is bewailing the wreck and praising those known to have been lost, almost nothing is said of the truly great philanthropist, Isidore Strauss of New York. If I were called on to pick out from the list the name of the one I thought will be most missed, I would certainly say Isidore Strauss, the man who fed the hungry and sold pure milk to the poor of New York at such prices as caused him to lose thousands of dollars and save thousands of lives. He sold these things in order that the poor purchaser might get them at such prices as they could afford to pay, and preserve their self respect. May God reward him! and raise up others to take his place.

May 23. I performed the marriage ceremony for white people the first time last night, uniting in marriage Mr. John M. Rogers and Miss Blanche Dunham, in Darien. The bridegroom is Railroad agent of Crescent and he and the bride, I think, reside in Liberty Co. above Ludowici on the G.C. & P. Ry.

June 4. We have just been informed of the death of Capt. Wm. Henry Atwood— he had gone on a visit to see his sons in Waycross, Ga. and went to bed in health on Monday night and was found dead this morning, it is supposed from heart trouble. I am told that his remains will be buried near Shell Bluff [Valona] tomorrow. He is believed to have been the last surviving commissioned officer of the Fifth Georgia Cavalry Regiment in the War Between the States. He was a fine old gentleman, 76 years old, of the old southern school.

July 16. We had a storm of considerable force from East all yesterday and last night. A part of the time the wind blew over 40 miles an hour and rain fell in torrents. The tides were unusually high. Some damage was done, mainly to rice on Butlers Island, where I am told there were several breaks.[214]

July 18. A white boy named Ray Gentry was accidentally drowned off the Black Island bridge about 3 o'clock today, and there is considerable excitement in our neighborhood on account of it. He was about 12 years old. He was crabbing and a loose board tilted with him, throwing him into the creek.

July 30. The first spike was driven in the extension of the Georgia Coast & Piedmont R.R. at Darien at midday today. It now looks very much as though they mean business.[215]

Aug. 12. Simon W. Devereaux, a highly-esteemed colored man, a citizen of Darien, died of heart trouble night before last and is being buried in the City Cemetery this afternoon.

Sept. 18. The machinery experiment for the Darien Ice & Light Co. has arrived and is being unloaded, so that it seems to be an assured fact that Darien is to have Electric Lights. We have now had telephones for about a year.

Oct. 14. The Darien Ice & Light Co. are still hard at work on their electric lighting plant and are hoping to have same in working condition in 30 days.

The G.C. & P. R.R. are [sic] still at work—it seems unintelligently—hoping to have their trains running to Brunswick within a year. But it seems to me that their hopes are not well founded. I do not understand them.

Oct. 21. We were very much shocked to hear of the death of J. R. Atwood (better known as "Tony") in Waycross last Saturday. He died of pneumonia; his remains were brought to Valona for burial yesterday.

Oct. 31. Mrs. Emma E. Strain died of heart failure at or about sunset today, which is quite a shock to the community. She was the earliest friend we had in this community and one of our most esteemed friends, but lately she became estranged from us, which we greatly regretted.

Dec. 2. Jos. Mansfield died this evening of a complication of diseases at his home in Darien, aged 70 years.

1913

Jan. 1. Christmas Day was fair, so is today—with these two exceptions rain has fallen since Dec. 22nd last. 1912 was noted for its extreme wetness, excessive warmth and its almost absence of breezes and winds. I have never seen so still a year. It was the wettest since 1895. We are entitled to about 52 inches of rain a year. There was an excess [this year] of about 13 inches. The largest excess I remember was that of 1895, which was 15 inches.

Feby. 1. The streets of Darien are now being lighted with electricity. Dwellings were so illuminated about a month ago.

The Brunswick extension of the Georgia Coast & Piedmont Railroad has now been finished to the river at each end from Brunswick and at Darien, and the trestles are being built.

March 4. I have been given the convict gang to overlook by the County Commissioners and will be allowed $20.00 extra compensation for March, April and May. I am to visit the camp once a week and must keep the city office open and pay all of my expenses out of the $20.00 paid me. I have gotten Mr. W. deR. Barclay to keep the office open when I am away and will visit the camp near Eulonia going by train and riding my bicycle back home.

April 29. The Georgia Coast & Piedmont R.R. is making good progress in their extension to Brunswick. We hope the road will be finished this year.

Railroad trestle over Darien River under construction, 1913.

May 11. Mr. John M. Fisher died at 6.30 p.m. Wednesday, aged 80 years, at his home on the Shell road of a complication of diseases, of which rheumatism was the more prominent.

July 4. I was re-elected Clerk and treasurer of Darien on July 2nd for the 11th time and I was elected Clerk of the Board of County Commissioners on July 1st for the 9th time.

July 16. Margie, Houstoun and Charlotte and Emily Britt go to St. Simons Island today to spend a week. Emma W. was to have gone too, but baby Claudia has fever today so her visit must be deferred. Claudie Britt will also go later.

Aug. 29. The City of Darien is erecting a water tank and tower—tank to be of 50,000 gallons capacity, placed on a steel tower 60 feet high.

Emanuel Hackel[216] has built a fine residence and is planning to build a store. Bluesteins & Downey are planning to erect a saw & planing mill so we hope Darien will now take on new life.

The Railroad across the river is not yet completed, but it is expected that it will be in the near future. The trestling is finished to Darien river and only the Iron Draw bridge over Darien river remains to be put in and the rail laid from Dents place [Hofwyl-Broadfield] to the Darien bridge.

Sept. 15. Rev. N. Keff Smith, now of Beaufort, S.C., is here visiting us.

Oct. 10. The City of Darien is erecting a 50,000 gallon water tank on a tower 60 feet high right in front of the Court House. This move is very unpopular because the citizens have been getting all the water they need for $2.00 pr. year heretofore but now they must pay from $6 to $10.00 pr. year and the people generally do not like that.

Oct. 21. I should have noted earlier that the City Jailers residence and the City Stables & store house were burnt on the morning of Saturday, 18th inst., this fire seemingly being the work of an incendiary—my theory being that it was set to spite the City authorities because of the erection of the water tank and tower.

Oct. 31. A dreadful accident. Alex Bailey's automobile "turned turtle" with him on the Shell road yesterday evening, injuring him so badly that he died at 2 o'clock this morning. There were two negroes in the machine with him. Both were hurt, but not dangerously. They were running the auto fully 40 miles an hour when the accident happened, but what caused the accident has not as yet been determined, perhaps never will be. The machine was badly wrecked. His remains will be buried in St. Andrews Cemetery tomorrow morning.

Dec. 10. A very bad shooting scrape occurred today—Mr. John P. Mew and his brother-in-law Edwin Sinclair—a boy—went to Generals Island to look after Mr. Mews hogs. There he found a negro man with one of his pigs. In the altercation that followed the negro man shot Mr. Mew in the face and head, shooting out, it is heard, both of his eyes. Edwin then grabbed up a rifle lying in the boat and shot the negro, killing him instantly.

Later—It is thought to have been likely that Mew, and not Edwin Sinclair, killed the negro. It is now hoped that at least one of Mews eyes can be saved, in an impaired condition enabling him to see to some extent.

1914

Thursday, Jan. 1st, 1914. Last year—1913—is in many respects a remarkable year. First, it is the dryest. Next, the warmest year on my records, or in my recollection.

Then, during this year there have been a greater number of improvements in Darien than in 10 years past. Water works were installed, electricity put in use, several new

Brunswick to Darien rail link completed, first train arrives, March 1914.

buildings have been put up, and the Brunswick extension of the Georgia Coast & Piedmont Railroad was built, except the bridge across Darien river—not yet finished—and they are building their new freight & passenger depot on Broad Street.[217]

Feb. 14. Rev. F. M. Mann,[218] colored Episcopal minister and ex-Postmaster, died in Savannah today, and will be brought here for burial next Wednesday. He was at one time highly esteemed by the white people.

Feb. 17. Alonzo Guyton, a very estimable colored man—for many years Deputy Marshal of Darien and a Democrat in politics—died last night at his home in Darien of Brights disease. His funeral will take place tomorrow.

Feb. 18. Rev. F. M. Mann and Alonzo Guyton were both buried this afternoon, Mann at 2 p.m. and Guyton at 3 p.m., both large funerals—and the strangest part is—Guytons sister, who came to his funeral, died of haemorrhage at Guytons home Tuesday night and so there were two corpses in the house at one time. Her remains were taken off on the Str. *Hessie* today.

Monday, March 2, 1914. This day marks an epoch in the history of Darien. The Georgia Coast & Piedmont Railroad ran their first passenger train into Darien from Brunswick carrying 3 coaches full of Brunswick business men up the road to Collins.[219] The bridge over the Darien branch of the Altamaha was made possible by working all day yesterday—Sunday.

March 9. Mr. August Schmidt died at his home in Darien on yesterday of paralysis after an illness of several weeks. He was about 78 years old. His remains were buried in St. Andrews Cemetery next day.

March 12. I went to Savannah yesterday to buy an automobile and returned last night, making the trip in about 6 hours not including stops. We left Savannah at 6.05 p.m. and arrived home at 1 a.m. We stuck in the mud in Bryan County once and stalled in the sand in Liberty County once and stopped at Eulonia about a half hour waiting on Dr. Clark who we met up with there. We wanted him to go ahead of us as he had good head lights and we have poor ones.

I bought a Ford 1911 from Dr. Crewther for $215.00 and my expenses of bringing the machine down was about $15.00 more. The machine is a 2 seated one—#38164 and is in excellent condition, except a few repairs that will not cost much. I will install electric lights on it which will cost about $10.00 extra.

April 23. Hugh Boyd Manson and Miss Sadie Clark[220] were married at the Methodist Church on the Ridge this evening by Rev. Thomas H. Thomson, who married the brides aunt. The entire community is in a stir over this event.

May 21. Hon. Geo. E. Atwood[221] died at his home at Valona of acute Brights disease this evening at 6.30 o'clock, aged 64 years. His remains were [to be] buried near his home. He was the Representative in the Legislature from this county at the time of his death and he and I served several terms together as County Commissioners some years ago.

Aug. 3. The most fearful war probably that the world has experienced has been commenced. Last Monday Austria declared war against Servia [sic], on Saturday last Germany practically declared war against Russia, and on yesterday, without a declaration of war, Germany invaded France. Today we hear that the navies of Germany and Great Britain had a fight, in which the British lost 2 vessels and the Germans lost 7 vessels. We hope this telegraphic news is exaggerated, but fear it is true. If so, we have Austria-Hungary-Germany, and I suppose Italy, arrayed against Russia, France, Britain and Servia—so far. Where it will end, no one knows.

The Exchanges the world over, even in this country, have closed and general business stagnation seems to stare us in the face—at least for the present. I cannot help wondering whether the battle of Armageddon is not about to be fought. If so, then Christ's second coming is not far away.

Aug. 5. The report of the naval battle mentioned above is untrue—But England has now declared war against Germany. Italy will remain neutral. No battles have been fought so far, but things look "as blue as indigo."

Sept. 20. My two grand-daughters, Charlotte Legare Britt and Emily Wilkinson Britt, joined the Darien Presbyterian Church today, for which I am thankful.[222]

Nov. 23. Hughie Timmons (Col), probably the oldest man in the community, died today in Darien of old age. He was about 100 years old.

Dec. 29. The weather is as dreary as I have ever seen, and too dark to work well. A fine mist has fallen all this morning so that everything is soaking wet. From a business standpoint this has been one of the dullest Christmases in 40 years here.

The war in Europe has now been fought to a standstill. Both sides have entrenched themselves and siege operations are carried on, one day in favor of Germany, next in favor of the Franco-British forces in France and Belgium, and one day the Russians get the better of it in the East, and the next day the odds seem to favor the Austro-Germans. It is estimated that 200,000 men have been killed or wounded, the Germans being the heaviest losers. The Germans have behaved like their Goth and Vandal ancestors and have repudiated their treaty obligations wherever this course has been of even temporary advantage to them.

1915

Jan. 28. I bought today in Savannah from Mrs. W. A. Sturtivant, for wife—one Model "T" 1910 Ford Touring Car #6620 (Shop Number) for $150.00. I then bought supplies for it that cost $14.00 and the expenses of Houstoun and myself in going for it were $5.00 more, making the $169.00 net cost. We found the Bryan County roads in very bad condition, and also a part of the Stage Road from South Newport bridge to Riceboro. The car is worth $100.00 more than we paid for it.

Feb. 9. Capt. Matthew J. Dean[223]—Pilot— died very suddenly yesterday evening at his home on the Ridge of apoplexia.

Feb. 10. I saw a strange sight this morning. I have been in Darien over 37 years and I saw for the first time today a bale of cotton sold in Darien. It was brought in on an ox cart and sold to R. A. Strain.

March 6. Seven white men were killed and 23 others badly wounded by a supposed madman named Marion Phillips in Brunswick today. Phillips was then killed by ex-Mayor Butts. His first victim, H. F. Dunwody, was raised here. This is the most awful havoc that I remember having been done by one man in so short a time—say 15 minutes. It is said that Phillips started out to kill 50 people and he came near doing it.

May 8. I was sent to Savannah yesterday by auto by the Board of County Commissioners to attend the Dixie Highway Convention there. I returned last night and enjoyed the trip.

May 11. The great ship "Lucitania" [sic] was torpedoed and sunk by a German submarine boat last Friday, and it sunk in about 20 minutes carrying down about 1500 passengers who were lost and the world—except Germany—is horrified that a so-called civilized country should be guilty of so horrible an offence against the laws of war. It is not yet known what the U.S. will do about it—about 100 of her citizens having been lost. It looks as if Germany is determined to draw this country into the present war!

Steel swingspan bridge over Darien River, 1914 (site of present US 17 bridge).

June 23. There is a "Basket Picnic" being held on the square between the Court House and the City Hall today. There is quite a large turnout of people from the surrounding country. The Brunswick Brass Band is here and there are several speakers on agricultural subjects here. Then a Baseball game &c., &c.

July 7. Reuben K. Walker, my old friend and neighbor, died very suddenly of apoplexy in his office in Savannah today, and his remains will be buried near Crescent.

Aug. 9. James Prindle attempted to escape from the chain gang on last Saturday, and was shot. He died in jail here today.

Aug. 31. We are still having a great deal of rain. The weather has been very hot, but is somewhat more pleasant now. The Rev. N. K. Smith is here with us. The Convict Camp is moved to Townsend today.

Sept. 4. Parson Smith is still here and is quite sick with appendicitis.

Sept. 8. Mr. Smith is still sick, but better. Rev. W. S. Milne is here on a visit.

Oct. 12. Savannah Presbytery will meet in the Presbyterian Church in Darien tonight. Presbytery was invited here against the wishes of a large majority of our people and I dread it; but must do the best I can to entertain the members for the honor of the church and community.

Oct. 14. Presbytery has adjourned. The attendance was small, there being only 18 ministers and representatives in attendance; this being so, we had no difficulty in entertaining them, so everything passed off pleasantly and, I hope, profitably.

Nov. 26. We observed yesterday as Thanksgiving day and as the 42nd anniversary of our marriage—we had the Whitesides family and Mr. and Mrs. Wm. Beckett to spend the day with us.

Nov. 30. Mrs. Charles O. S. Mallard, one of our Elders, died at the home of his daughter Lilla, Mrs. J. T. Howe, in Kinderton, Ga. at 5 a.m. yesterday and was buried in St. Andrews Cemetery here today.

1916

Jan. 6. Houstoun got dreadfully injured today. He was returning from home where he had been to dinner, about 2 o'clock p.m., and ran into a Ford automobile being driven by Mr. J. G. Forbes[224]—through no fault of Houstouns—just west of the bend in the Shell road, he was riding a motorcycle. The impact of the machines was so great that the automobile was turned completely over. Mr. Forbes escaped serious injury, but Houstoun had his right thigh broken in two places and his nose was broken and his head and face cut and badly bruised. We took him to Brunswick Hospital at once by special train, and then had his broken leg set and his wounds dressed, and I was much relieved and very thankful when the physician pronounced him not fatally hurt, for we thought him fatally injured.

Jan. 11. Houstoun has improved greatly. I took his mother and Emma over on the Str. *Hessie* yesterday to see him.

There was another auto accident here today, W. A. Brannen[225] coming out the Kells Grove Avenue into the Shell road, ran into Dr. [P. S.] Clark's auto, bruising Capt. Jas. Lachlison's hand and face and breaking the front wheel to Brannen's car and doing some damage to Dr. Clark's car.

The accident in which Houstoun was injured caused a greater commotion in Darien than did the riot of 1899. Everyone seems to have taken it to heart. I am thankful that

Houstoun was in no way to be blamed for the accident.

Feb. 25. Mr. Porter Middleton[226] is building a brick store on Broad Street on the site of the old store of Collat Bros. next [to] P. S. Clarke & Co. Drugstore.[227]

March 16. I took a trip by auto to Chatham County today, taking Mr. F. H. MacFarland, Geo. W. Poppell[228] and Houstoun with me. We went with a view to investigating the methods of using and working their convicts and to see what was being done to the roads, through which the Stage road passes.[229] We visited the camps and gangs of Liberty, Bryan and Chatham counties and returned this afternoon.

April 29. Houstoun has so far recovered from injuries received Jan. 6th that he can now get around with the aid of a walking stick. He is at work again, having been given charge of the Ice and Electric Departments of the Darien Mfg. Co. on April 23rd, their late manager, William Beckitt having left them.

June 8. At a meeting of the new City Council held yesterday evening Jesse A. Britt was elected Mayor of Darien. The new Aldermen are R. A. Young, J. A. Space[230] and A. P. Lee[231] with one other to be chosen. E. G. Cain, Jr.[232] was elected an Alderman but refused to qualify.

June 15. I am being fought stubbornly for the position of City Clerk by J. G. Forbes, A. P. Lee, R. A. Young, Mr. Robert Manson, and others not so prominent.[233] They have succeeded in cutting the salary of Clerk from $50.00 to $35.00 and want to put Raymond W. Clancy[234] into the office. There was a preliminary fight at a meeting on yesterday for a fifth member of Council. Messrs. D. S. Walker[235] and Jim L. Stebbins[236] were put up—and Jim Stebbins elected on the second ballot. He has promised to vote for me, and I hope he will, at the July meeting.

Mr. Forbes—it will be remembered— is the person who caused Houstoun to be so dreadfully hurt on Jan. 6th, 1916, and he seems to be fighting me because he does not seem to think he injured us enough then! But "God sees," and "He has delivered me hitherto, yea and will deliver me…"

July 13. I was re-elected Clerk and Treasurer of Darien for the 13th time on July 5th, without formal opposition, the fight on me having been dropped when J. L. Stebbins was elected Alderman.

Oct. 15. Mrs. Geo. Patelidas died at the City Hospital in Brunswick, Ga. on Friday of Blood poisoning—after a long illness. Her remains will be buried in Brunswick this afternoon. She left a young baby and a little son about 2 years old, besides her husband who keeps a restaurant and fruit store in Darien and is well thought of. Very sad—

Dec. 23. Two fires in town! Last night a house near the old G.C. & P. R.R. [at Columbus Square] belonging to M. Bluestein and occupied by Mr. Sparkman came near burning down. This morning the Presbyterian Manse caught in the attic and only quick work by the "bucket brigade" saved it.

1917

Jan. 11. I have just received new auto tags for 1917—they are for my car #2509 [and] for Mrs. Legares car #2552.

On last Tuesday I was re-elected Clerk of the Board of County Commissioners for the term of the Board—2 years, and perhaps 4 years—for which I am very thankful.

March 13. Since the last entry a great many important things have happened that ought to be recorded, some of which are—The U.S. has broken off diplomatic relations with Germany because of its ruthless conduct of its submarine warfare so that our country is now on the verge of war. The English have captured Baghdad from the Turks and are making progress in their invasion of Palestine. Does this foreshadow the great Har-Miggiddo battle on the plains of Eschaelon? I am inclined to think so. The British are also driving the Germans from their positions in northern France. This is we hope "the beginning of the end" of this terrible war.

As a result of the war, prices have been advanced in America unreasonably. Flour is sitting at $10.00 pr. barrel wholesale. Onions are selling at from 2 to 5 cents each retail. Irish potatoes are $10.00 pr. 100 lb. wholesale. Meats are very high, so are canned vegetables. Cabbages are not to be had. Shoes going up steadily. Corn $1.35 pr. bushel, oats 80 cents wholesale in Savannah. My impression is that speculators are responsible for the awful prices.

March 30. The Legislature has passed the "Bone dry" prohibition law, which the Governor signed at midnight Wednesday, which makes anyone having a drink of liquor in hand guilty of a misdemeanor—thus making probably three fourths of the good people of Georgia legal delinquents on Thursday morning. I do not know just what to make of the Governor and Legislature. But I do know that the law will not work here.

April 6. Congress on yesterday declared that because of the ill treatment of American vessels by the Germans, "That a state of war exists between the U.S. and Germany" so that we are into the war against our wills.

May 8. My old friend Capt. A. S. Barnwell died in Savannah yesterday—will be buried in Bonaventure Cemetery. He was 84 years old.

Railroad automobile transfer over Altamaha delta, Darien to Broadfield, ca. 1916.

June 7. The Presbyterian Sunday School has given a basket picnic at the Hopkins place[237] at Crescent today, to which the other Sunday Schools and the public generally were invited. The G.C. & P. R.R. gave a reduced rate of 35 cents for the round trip, but the train was 2 hours late.

Sept. 19. Note. I will enter here the result of my investigation into the time of the birth of Christ—the true Christmas. History tells us that Tiberius assended [sic] the

Roman throne Aug. 18, A.D. 14. Luke tells us (Chap. 3, Vs. 1) that John the Baptist commenced his ministry in the 15th year of Tiberius, probably in April when he became 30 years old. Luke tells us (Chap. 1, Verse 26) that our Lord was 6 months younger than John. John must have been 30 years old in April, A.D. 29. It is generally admitted that our Lord's ministry continued 3 1/2 years—or until he was 33 1/2 years old. It is probably correct that he was crucified at 9 o'clock a.m. and died at 3 o'clock p.m. April 15th. But, what year? Most scholars say A.D. 30. I think the Bible would say A.D. 34. It is therefore probable that our Lord was born on or about Oct. 1-5 B.C. 1 or, 3 months B.C. of the annum [illegible]. These dates are from the *Bible*, not Josephus, so that Old Dionisius Exeguus was right in his calculation of the date of the birth of Christ, and this is correctly 1917.

I desire to call attention to the fact that the Lord was probably born on the Jewish new year, the Feast of Trumpets, and that that is the day we ought to expect His return to earth at His second coming—which will be in the near future. God grant that all mine may be ready to welcome him!

Dec. 18. A meeting of citizens was held here yesterday having in view the establishment of a shipbuilding plant here. The outlook is very favorable except that nearly all of the preliminary arrangements have been made on Sunday, which means that it must proceed without God's blessing on it, which is unfortunate. Some enterprise of this kind *must* come into this community, or we might as well pack up and move away—all of us—for business is very nearly dead in town.

1918

Jan. 3. The Darien Shipbuilding Company met at the City Hall at midday today and organised. They elected their Directors and officers and will go to work very soon, I am told. I certainly hope so. Robert Manson is president and F. H. MacFarland is Vice President and R. J. Downey is Secretary and Treasurer.[238]

Jan. 12. The fuel situation is serious all over the world. Steamer *Hessie* has not left Brunswick for Darien since last Wednesday—can't get fuel[239]—The N.Y. steamer for Brunswick is lying up in New York—can't get coal, and there is a shortage of a great many articles in Darien, due to this cause. The railroad trains are running, but out of schedule.

Jan. 14. The Darien Shipbuilding Company began work on their shipyard in Darien today, and we are rejoiced that something has turned up to enliven business here.

Jan. 23. This is the 40th anniversary of the arrival of myself and family to live in Darien.

June 10. Houstoun and Miss Lucilla (Dollie) Wells[240] were married quite unexpectedly last night at her home near Mt. Vernon in Montgomery Co., Ga. They returned to my home this evening. They will stay with us for the present. We are pleased with the match and wish them much happiness.

June 12. Licentiate John C. Blackburn has accepted the call to the Pastorate of our church and has arrived in the field and taken charge. He is engaged to preach for us each 1st and 3rd Sunday at a salary of $45.00 per month and a house to live in.

Aug. 23. Richard W. Grubb, Editor of the *Darien Gazette*, died at a hospital in Savannah yesterday evening of severe bladder, and possibly kidney, trouble, the immediate cause of death being toxhaenia (blood poisoning), aged 66 years. His remains have been brought here and his funeral will be held in St. Andrews Episcopal Church of which he was a Warden. He was a good man and an excellent citizen and will be greatly missed. His paper, the *Gazette*, has not been published since Aug. 3rd and, at this writing, it is uncertain when it will re-appear.

Oct. 10. The war is progressing favorably for the Allied armies and it is hoped that the Germans will be driven out of France before winter.

Oct. 29. Our shipyard here obtained a contract from the U.S. government to build two large barges. Now we are told that the contract has been annulled. Poor Darien. The town has been dead for some time. This lays her out. The next disaster will bury her, I fear. And the worst of the matter is that its citizens do not realize that the town is dead!

In war news, there is nothing startling, but signs point to an early surrender by the German people to the Allies. Turkey is about conquered; Bulgaria has quit; and now Austria is offering to capitulate, and the Germans are being slowly forced out of France and Belgium.

Nov. 11. We have received reliable information that Germany has accepted and signed the Armistice offered them by Gen. Foch representing the Allied and U.S. governments, thus bringing the great war to an end, and the community [is] celebrating the event by ringing the bells and making all the noise they can. The schools have been dismissed and the churches will hold thanksgiving services.

Street scene, Townsend, western McIntosh County, (C.H. Davis-E.M. Thorpe store).

1919

Jan. 9. On the 7th Jan. I was re-elected Clerk, B.C.C., for 2 years and *after my election and acceptance of the office, my salary was reduced from $70 to $50 per month!* I do not think this act is legal.

Feb. 14. My first Legare grand-son, son of Houstoun and Dollie, was born at 10.20 a.m. today [see note 240].

June 13. Unusual—We are in the midst of a N.E. gale of about 30 miles an hour, accompanied by heavy rain! It rains heavily every day now and is cloudy nearly all the time.

This is a peculiar season. There were no plums, nor black beans, very few peaches, about a half crop of figs. Vegetables are almost a failure due to the protracted drought in April and May. Labor is not to be had at any price—the negroes will not work. What is up I do not know but I suppose we will find out soon. Negro men want $2.00 to $3.00 per day and one cannot get them at that. The negro women are giving up their jobs—no wash women to be had at any reasonable price.

A general spirit of unrest seems to pervade the whole world—what does it portend? Is this the "End of the Age" foretold by Daniel and John? If so, I say "Amen, come Lord Jesus!"

Aug. 20. I took a trip by auto to Brunswick today to have a set of lower teeth made.[241] I find it will cost $35.00 to have 5 teeth bridged in.

Aug. 28. The passenger train on the G.C. & P. Ry. broke through the Altamaha trestle just south of the bridge over Darien River today. No one was seriously hurt. The engine tender and one box car of lumber fell to the ground. The engine is still on the trestle, but may fall any time unless steps are taken to bolster it up.

Sept. 1. Death—Old Mrs. Roughin[242] died yesterday at the Sapelo Light house. She was [blank space] years old and died of [blank space]. Her remains will be buried in St. Andrews Cemetery some time today.

Sept. 8. Fire in town—A fire of misterious [sic] origin destroyed the building used as the Telephone Exchange last evening. The manager, and his family lived in the building.[243] They had gone out for a visit leaving no fire in the building. They returned at 8.30 p.m., not quite dark, and found the building on fire and the roof about to fall in. They lost every thing they owned except a baby carriage. The building was owned by W. C. Wylly and was uninsured.

Sept. 10. On Aug. 28th the passenger train on the G.C. & P. R.R. broke through the trestle on Generals Island right near Darien and traffic of all kinds has been suspended entirely until this morning when traffic was resumed. They got the engine on the track yesterday evening.

We have just received the distressing news of the death of Dr. P. S. Clark, our old friend and physician (the only physician nearer than Dr. Long at Townsend) at a hospital in Savannah. He was hurridly taken there last night and died at midday today. He had a bilious attack a few days ago, but Dr. W. J. Long[244] did not think him dangerously sick when I talked with him. We will sadly miss him. He had his faults, but he was a good man!

Sept. 12. The remains of Dr. Clark were buried (from the Ridge Methodist Church) in St. Andrews Cemetery at 11 a.m. today, his being the largest funeral I have seen since my arrival here 42 years ago. I cannot figure out how we will get along without him.

Sept. 27. Mr. William B. Hagan[245] died at the residence of his son S. J. Hagan[246] in Darien yesterday evening. He was 78 years old and died from a stroke of paralysis.

Handle factory at Kell's Landing, Lower Bluff, near Darien, ca. 1917.

Oct. 20. We have just heard that federal Judge Evans has confirmed a sale made of the Georgia Coast & Piedmont Railroad to junk dealers for $426,200.00 and that the road will probably be torn up and junked after Nov. 12, 1919–This is another calamity for Darien.

Oct. 22. I went down the road to Union Island[247] today in my launch, taking wife, Emma Whitesides and the Misses Anne Lee and Frances Haynes.[248] We broke down and the girls had to row the most of the way home, but in spite of this mishap they seemed to enjoy the trip. I worked with the engine until I was completely exhausted.

Mr. Thos. A. Bailey, who has been ill in a hospital in Savannah, died this morning.

Oct. 23. Thos. A. Baileys funeral was held at his home on the Ridge today and the burial took place in St. Andrews Cemetery immediately after. He was one of my oldest friends. He died of a complication of diseases—mostly cancerous—He was Chairman of the County Commissioners. I was one of the pallbearers. He was 66 years old. He was an upright man.

Oct. 31. Death—Chas. L. Bass[249] dropped dead on Sapelo Island yesterday from heart trouble. His remains will be brought to Darien for funeral & burial today in St. Andrews cemetery. My old friends seem to be dropping off like the autumn leaves.

Nov. 11. This is "Armistice Day" being kept as a [illegible] holiday.

The Georgia Coast & Piedmont railroad has been sold to junk dealers—to be delivered to them today—No trains are being run. We are uncertain as to what will become of the road—whether it will be taken up or not. We do not know yet whether the ferry will be operated, but I think it will be as it is a good-paying institution.—Poor Old Darien! Within the last month about 5 families have left the community—and others must follow if the road is discontinued. The town is *dead.* The only question to be determined is how deep it will be buried. But, God sees!

Nov. 24. Fire—More trouble for Darien—About one half of the plant of the Altamaha Woodworking Co. was burnt last Saturday night.[250] This burnt part was the "Hardwood factory" where hickory and other hard woods were worked and was the most valuable and useful part of their plant. Their hardwood warehouse contained a large amount of manufactured axhandles, [and] spokes & was also destroyed. I have no idea that these buildings will be rebuilt.

These times remind me of the years 1863-64 when we had bad news almost continuously, and of 1865-73 when times were harder than they are now but, by Gods help, we lived through them.

Dec. 26. We have just passed probably the quietest Christmas passed in these parts since 1865. There is almost no noise.

1920

Jan. 1. Some observations on the year 1919. This year has been one of hardship and disaster. It was a poor crop year. Everything that I can think of except salaries and sugar increased greatly in value. My salary was cut from $70.00 to $50.00 per month and my work about doubled. But we are thankful that we are as well off as we are. Two of our foremost citizens have died. Our railroad has been "junked." Darien continues to shrink up with no immediate prospect of revival.

I am told that several Darien merchants will go out of business and leave the town. Houstoun has lost his job here and must go elsewhere to work. Wife's health continues very poor with no signs of improvement.

Jan. 12. Mrs. Catherine Malcolm[251] died on Sapelo Island yesterday morning and was buried in St. Andrews Cemetery today. She was 95 years old.

Jan. 20. Houstoun, Dollie & the baby arrived here from Montgomery Co. last night. They will remain with us a few days then he will go to Dublin, Ga. where he has been employed as Engineer of their Light & Water works—at least for the present. We regret that he must go so far away, but are thankful that he has found employment.

It is generally reported, and I believe it, that Miss Jean Manson and Dr. Charles C. Fishburne[252] were married in Savannah today—it seems like a "Runaway match" made with the consent of the parents and in their presence.

The Board of County Commissioners at their meeting Jan. 6th restored my salary to $70.00 per month.

Jan. 28. Houstoun has notified us that he refused to take the 2nd Engineers job offered him at Dublin because they required him to enter into contract for stay for several years. We do not know what his plans are for the future.

Jan. 31. The wind is blowing a gale of about 40 miles per hour from N.E. and the weather is pretty cloudy, not cold. Several vessels have been lost on our coast since the N.E. gale commenced last Sunday—one of them a Spanish vessel loaded with lumber is now going to pieces off Sapelo Island about 4 miles north of the Light House. It is not known what has become of the crew.

Hands are at work tearing up the railroad just north of the Shell road, the road having been destroyed from Ludowici that far towards Brunswick.

Feb. 5. The wreckers of the Georgia Coast & Piedmont railroad took up the rails as far as the river today—Poor old Darien! It surely seems to be a hopeless wreck.

March 30. On March 20th I formally entered the race for Tax Collector of McIntosh County. The Primary Election takes place on April 20th. My opponents are S. J. Hagan & W. C. Wylly.

April 21. The Election returns are now all in. I received only 20 votes out of 189 cast, but am satisfied with the result, except that it has given me a very poor opinion of some of my *friends*, who of their own unsolicited free will, promised me their votes. I do not expect to ever again become a candidate for any political office, except those I now hold which may be considered semi-political.

April 27. I celebrated my 68th birthday yesterday, quietly at home, not being at all well. Today is Mother's birthday. She would have been 89 years old had she lived.

May 8. Mr. A. C. Terrell, while returning from his daughters—Mrs. Ivey Ryals—on Black Island to his home near Townsend, took passage as far as Darien with Britt in his auto and died on the way in of heart failure. He seemed to be in good health before the sudden attack—no physician in the entire community, at least nearer than Townsend.253

June 5. I was re-elected City Clerk & Treasurer last Wednesday for the 17th yearly term. Within the last week the City has installed a 35 H.P. oil engine and bought the electric equipment of the Darien Mfg. Co. at a cost of about $13,700.00 which I do not think can be paid. Our authorities do not realize how nearly dead the town is—we have lost about 600 inhabitants since the census of 1910, and the houses are going to ruin, the most of them.

Vernon Square, Darien, looking southwest toward delta (Methodist Church at right).

June 9. The County Sunday School Association is having a picnic at Eulonia today. The Britts, Whitesides and Margie have gone to it.

July 1. John C. Blackburn resigned the pastorate of the Darien Presbyterian Church and left for Douglas, Ga. June 23rd. Houstoun has accepted a position with the DeLoach Quick Ice Mch. Co. of Columbia, S.C.—he left June 22nd.

Aug. 24. We have had no rain since the last entry [August 11, 1920] until last night, when we had a thunderstorm and about 1 in. of rain. Lightning struck the steeple of the Negro Baptist Church and damaged it so much that the whole top part will have to be rebuilt.

Emma Whitesides and her children left today by Str. *Hessie* for a visit to friends in Cochran, Ga.

Sept. 10. I accompanied the Chairman B.C.C.[254] to Brunswick today for the purpose of attesting the agreement between Glynn and McIntosh counties under which they will purchase the bridges and rights of way of the defunct Georgia Coast & Piedmont railroad for the purpose of turning same over to the State Highway Commission to be made a Public Road between Darien and Brunswick. By invitation we dined with the Young Mens Club at the Oglethorpe Hotel. We were most hospitably entertained.

After dinner we attended a meeting of the Glynn Co. Commissioners and managed our business satisfactorily and returned home that evening.

Nov. 3. Capt. James Lachlison, 83 years old, and one of our oldest and most respected citizens, died Nov. 1st and was buried today.

1921

Jan. 27. The State Highway Commission is building a highway over the road bed of the old G.C. & P. R.R. across the Altamaha delta from Darien to Glynn Co. We hope the road will be finished by April 1st, 1921. They seem to be making a good job of it, but are getting along slowly.

Saturday, July 2nd, 1921. A Red Letter Day for Darien.[255] The Darien-Brunswick road was formally opened (tho not quite finished) today by suitable exercises. There was a barbecue in Darien—about 40 animals being barbecued. There was much speaking, the following being [among] the chief speakers—Gov. Hardwick,[256] and the Chairman of the State Highway Commission, Mr. Strachan, and the Chairman of the Dixie Highway[257] Commission, Judge Patterson.

There were probably 4000 people in Darien and about 1000 automobiles. The crowd was the most orderly of its size I ever saw. The day was partly cloudy and pleasant and the crowd seemed to have a good time. This road will —we hope— be of great benefit to this entire section of country.

Aug. 30. There have been several bad accidents on the road from here to Brunswick lately, all of them the result of carelessness—perhaps drunkeness. Last Sunday night a Brunswick truck filled with negro frolickers upset on Champneys Island killing a negro man named Dorsey, and breaking a womans leg.

As to the liquor question. The present prohibition laws seem to be a complete failure. Those who ought not to have it, but want it, have no difficulty in getting all they can pay for, but the stuff they get here is poisonous. I am told that those who have the money have no difficulty in getting good liquor in either Savannah or Brunswick, but it costs $8.00 a quart bottle, or 50 cts. a drink.

1922

Feb. 20. I have been ailing from sciatica[258] for some time—became so ill on the 9th that I had to give up and go to bed. I am out again today but am not well yet. This has been the most painful disease I ever had.

March 9. I took a trip to Savannah by automobile accompanied by Houstoun and

Steel railroad bridge boarded over for automobile traffic, looking north, ca. 1923.

Legare Britt and bought Ford Touring Car from Roy Chalker (second hand) for $230.00. I consider it a bargain. It is owned by wife.

April 5. Wife died this morning—She was taken suddenly very ill about 1.20 a.m. April 3rd with some heart trouble, becoming insensible and falling off the bed before I had time to help her. She had what I suppose to have been a spell of angina pectoris. She was very ill all Tuesday and Tuesday night, but was in possession of her senses. A few minutes before 6 a.m. Wednesday—this morning—she asked me for a drink of water, which I gave her; she took a little of it and before I could take the glass away from her mouth (I was supporting her in a reclining position) she commenced to breathe badly. I then laid her down and felt her pulse. She had none and stopped breathing in a few seconds, dying at 6 a.m.[259]

April 6. Wife was buried in my lot in St. Andrews Cemetery at 11 a.m. today. Many friends attended for which I am thankful. Rev. L. E. Williams[260] of the Methodist Church preached her funeral. The pallbearers were Messrs. C. M. Tyson,[261] W. deR. Barclay, E. G. Cain, Jr., R. K. Hopkins,[262] John Manson, and J. G. Forbes. The floral offerings were beautiful. I am more than ever like the "Lone sparrow on the house top" (Psalms 102.7). But it is well. Her children were all with her.

April 17. Capt. Thos. S. Wylly,[263] aged 91, died in New York on the 13th and was buried here in St. Andrews Cemetery today.

1923-1932

Sept. 25, 1923. Wyatt deR. Barclay, Mayor of Darien, died at about 2 o'clock p.m. after several months illness of heart disease. [Buried at St. Andrews Cemetery].

Oct. 9, 1923. William C. Wylly died at the Brunswick Hospital this morning of Brights disease. He was 82 years old and will be buried from the Episcopal Church in St. Andrews Cemetery.

Feb. 15, 1924. Rev. N. K. Smith, D.D. came yesterday to pay me a visit of several days.

Looking north on Dixie Highway (US 17), Altamaha delta near Darien, 1920s.

Feb. 18, 1924. Mrs. Geo. Atwoods[264] residence at Valona was entirely destroyed by fire about 3 o'clock p.m. yesterday. Some of the furniture was saved.

June 24, 1924. Yesterday afternoon I took a trip over the new Brunswick-St. Simons road with Mr. J. Van Dyck and enjoyed the trip very much.

July 11, 1924. Brunswick is having a big day of it. The road from Brunswick to St. Simons Island will be formally opened today with much speaking and many ceremonies, and I have a half-holiday, but will not go to the celebration as I don't like crowds. The Britts and Whitesides have gone.[265]

Oct. 29, 1926. In starting my car in front of the City offices it got away from me and rolled down a slight declivity then, and knocked me down and ran over my chest. I did not think much of it at first, but about 3 hours after was found in my office chair insensible by Emma Whitesides and others. Dr. Fishburne was sent for and found that two of my ribs were broken, one broken in two places. [Editor's note: the following was added later]. I was compelled to go to bed and stay there until Nov. 19th when I managed to go to the office and conduct a little of the most important business.

Nov. 23, 1926. I am not yet well, but feel much better today.

Jan. 11, 1927. A disastrous fire in town last night. Fishers 4 buildings, Mrs. Orrs house, the old Hilton & Dodge Lumber Co. office, and Strains store were entirely burnt. It is thought that the fire originated in the old Fisher bakery. The Brunswick fire department was here and did much good.

Feb. 9, 1927. I was relieved as Clerk by McIntosh County and by the City of Darien on Feb. 1st & 2nd, 1927, the County giving me a pension of $25.00 as long as I live, monthly.

March 14, 1928. There seems to be a building boom on in Darien. The following buildings have been rebuilt or in the process of being expanded—the Public Library, the old Rothschild store, the Tyson building where A. Konetzko had a 10 cents store and the Glynn Ice Co. is building a refrigerated [illegible] room on Lot 111 just north of the old Blacksmith shop.

Aug. 13, 1928. W. J. Hazzard[266] died in Brunswick of a complication of disease last week and was brought here for burial in St. Andrews Cemetery. Robert A. Strai died in Brunswick and was buried in St. Andrews Cemetery this evening.

Monday, Sept. 3, 1928. My daughter Margie Hart was taken suddenly ill at church yesterday and was taken home where she died at 7.15 p.m. and will be buried in St. Andrews Cemetery tomorrow afternoon. Her illness was paralysis of the brain. She was ill only about 7 hours.[267]

Nov. 5, 1929. The City of Darien is holding an election under the new charter for Mayor and Aldermen. The candidates are R. Austin Young and H. Pat Brannan[268] [for mayor]. Aldermen—M. Bluestein, E. G. Wilkins,[269] R. S. Townsend[270] and D. E. Lane.[271] Young received 105 votes and Brannan 86. The Aldermen had no opposition.

Nov. 13, 1929. Mr. George Corlette,[272] age 81, died of cancer sometime yesterday and was buried in St. Andrews Cemetery today.

Darien civic leaders, 1930, (l-r): Robert Young, Paul Varner, Legare Britt, Middleton Davis, John Clark, Gene Cain, K. S. Trowbridge.

Sept. 8, 1930. Two negroes, Fred Bryan and George Grant in an attempt to rob the Darien Bank last night about 12.30 shot and killed county police[man] Freeman of Glynn Co. and injured R. J. Anderson,[273] J. M. Fisher, Jr.[274] and Deputy Sheriff Collins. Anderson saw the negroes about the post office and hid near Britt's store. As the two negroes crossed toward the bank one of them spied Anderson and stopping over him shot him three times. Fortunately no one of the shots was serious. Anderson succeeded in chasing them down into [illegible] old rice field and sounded an alarm for help. The police from Brunswick arrived in a short time and went after the negroes who were hiding in the tall grass. Just as Freeman parted the grass one of the negroes shot him in the right eye and over the head. He died before they could get him to the Brunswick hospital. Collins, though wounded seriously, is not dangerously hurt while Fisher's wound was slight. Practically every man in Darien joined in the man hunt and early this morning Grant was captured and in spite of assistance from Savannah the criminal was shot upon reaching the jail. An airplane from Jacksonville arrived about 10 o'clock to take up the chase of the other negro. Up to the present he has not been captured. Considerable uneasiness is felt throughout the entire community as both negroes are from this section.

Sept. 15, 1930. Claudia Whitesides left this morning to commence her senior year at the State University in Athens, Ga.

March 24, 1931. Shortly after 2 o'clock a.m. the Court House was discovered on fire and for a short time it looked as if it would burn entirely down but the firemen had the fortune to put the fire out, but the whole Court House is badly damaged.[275] The records, however, were saved. The City Court for this County is in session and it is supposed that some of the jurors or witnesses threw a lighted cigarette or cigar in some saw dust under the stair and the saw dust took fire.

April 5, 1931. Easter Sunday. Rained several times today which I think will discourage the women from wearing their Easter finery. Lewis Massey, Charlotte and Virginia Massey are here.

Today is the 9th anniversary of wife's death. I have been unable to visit the grave, but I got Claudie to visit it for me. We had a delightful communion service today. Claudie returned to Waycross with Lewis and Charlotte for an extended visit.

Saturday, June 12, 1932. Jos. Whitesides surprised us by marrying a Glynn Co. girl last Tuesday night and we are only now [getting] the facts. He brought her over last night.

We have had several rains lately.[276]

[Editor's Note: On the following page after the above final entry in Legare's journal, his daughter, Emma Legare Whitesides, wrote the following]:
Darien, Georgia, Oct. 16th, 1932.

The diary of my father, John G. Legare, ends here.

He was taken sick on Oct. 8th but not seriously enough to cause alarm. He was much better Wednesday, but on Thursday he developed pneumonia and became rapidly worse. Dr. Ogden was called and he advised special nursing so Mary V. Cromley came to the case. Father was not able to resist the disease in spite of the splendid attention he received and on Friday night at 9.15 he died. Today, Sunday, Oct. 16th, he was laid to rest in St. Andrews Cemetery beside my mother, the Rev. F. M. Baldwin[277] preaching the funeral service. The active pallbearers were R. H. Manson,[278] R. W. Clancy, W. S. Tyson,[279] A. S. Poppell,[280] E. G. Wilkins and Edd Thompson.[281] Honorary pallbearers being C. M. Tyson, E. G. Cain, J. F. Thomson,[282] R. K. Hopkins, M. Bluestein, Wilkes Poppell.[283]

In spite of a terribly stormy day with rain coming down, nearly every family in the community was represented at the funeral and the many beautiful flowers testified to the high esteem in which my father was held—

Emma L. Whitesides

Notes

1. Nathanial Heyward Barnwell, proprietor of Evelyn plantation on the Altamaha delta in northern Glynn County, Georgia. Barnwell, who had come to the area from South Carolina, was evidently leasing the Evelyn tract for rice cultivation from its owners, the heirs of Hugh Fraser Grant, Sr. (1811-1873). Grant was one of the leading antebellum rice growers of Georgia, planting most of his crop on Elizafield, the adjoining tract to Evelyn. N. H. Barnwell acquired ownership of Evelyn in the 1890s. The Evelyn house, built in the 1830s, was the residence of Barnwell during the 1880s and 1890s. Barnwell also leased ricelands on Generals Island near Darien, Georgia, where John G. Legare was employed as manager from 1878 through 1896. Norman J. Cobb and Don McWaters, "Historical Report on Evelyn Plantation," *Georgia Historical Quarterly* 55 (Fall 1971), 424, 429. See also the Introduction to the present work.

2. Savannah's population in the 1870 U.S. Census was 28,235, an increase of about 6,000 inhabitants from the 1860 Census taken just prior to the Civil War. Roland M. Harper, "Development of Agriculture in Lower Georgia From 1850 to 1880," *Georgia Historical Quarterly* 6 (June 1922), 114, 117.

3. Sterling, Glynn County, Georgia, was located about ten miles west of Brunswick, the county seat. Sterling was a stop on the Macon & Brunswick Railroad. Before the opening of the Florida Central and Peninsular Railroad in McIntosh County in 1893, Sterling was the nearest rail access to the interior for Darien residents. Joseph Tillman, C. P. Goodyear, eds., *Southern Georgia: A Pamphlet Published Under the Auspices of the Savannah, Florida & Western Railway, Brunswick & Albany Rail Road and Macon & Brunswick Rail Road* (Savannah, Ga.: Savannah Times Steam Printing House, 1881).

4. Legare's family: wife, Charlotte Smith Hamilton Legare (1848-1922); stepson, Arthur St. Clair Hamilton (1868-1900); daughters, Claudia Girardeau Legare (1875-1962). Margaret Hart Legare (1878-1928) and Emma Strain Legare (1885-1973); son, James Houstoun Johnston Legare (1889-1965). See notes 125, 199 and 240 for extended family history.

5. Carteret's Point is south of Hofwyl-Broadfield plantation overlooking the north end of St. Simons Island.

6. General's Island, named for General Lachlan McIntosh who owned the tract prior to the American Revolution, is situated across the Darien River from the town of Darien and bounded on the south by Butler's Island. In 1853 Generals Island became one of the properties of the John and Pierce Butler Estate and a portion of the Butler's Island slave force was employed there to cultivate rice. Buddy Sullivan, *Early Days on the Georgia Tidewater, The Story of McIntosh County and Sapelo* (Darien, Ga.: McIntosh County Board of Commissioners, 5th edit., 1997); Buddy Sullivan, ed., "The Agricultural Journal of Roswell King, Jr., 1845-1854" (Unpub. ms., 1997); U.S. Coast Survey Topographical Map, "Altamaha Sound and Vicinity, Georgia," 1869.

7. See the Introduction to this volume for an overview of the importance of the timbering industry to Darien during this period. For a detailed review, see Sullivan, *Early Days on the Georgia Tidewater*.

8. Adam Strain (1840-1897), Darien merchant, in 1874 was proprietor of a gen-

eral store and ship chandlery specializing in groceries, dry goods and hardware located on the upper bluff of the Darien River on the southeast corner of Broad and Screven streets. At this writing in 1997, this two-story tabby-stucco structure was still standing. It is the oldest building in Darien, being constructed in 1815-1820. The exterior walls survived the torching of Darien by Union forces in 1863. Strain was one of McIntosh County's most prominent citizens, being a leading business and civic figure in Darien during the postbellum period. He was a McIntosh County Commissioner and president of the Darien Bank, 1889-1897. He is buried in St. Andrew's Cemetery near Darien. *Darien Timber Gazette,* May 16, 1874, June 13, 1879; Evelyn K. W. White, Index to McIntosh County Cemetery Records (Darien, Ga.: Unpub. ms., 1958), 42; Minutes of the Board of Directors of the Darien Bank, 1889; Buddy Sullivan, *The Darien Bank: A Celebration of 100 Years, 1889-1989* (Darien, Ga.: The Darien Bank, 1989).

9. Hiltons & Foster Lumber Company, later Hilton Timber & Lumber Company and, from 1888, the Hilton-Dodge Lumber Company (see also note 28). Hilton Family Papers, Collection 387, Box 1, Folder 5, Georgia Historical Society, Savannah, containing Thomas Hilton, *High Water on the Bar,* privately printed, 1951, cited hereinafter as Hilton, *High Water;* Box 1, Folder 3, Hilton-Dodge Lumber Company records.

10. Evelina Myers Blount (1813-1887) was the widow of Edmond M. Blount (1811-1866), prominent Darien citizen who held several McIntosh County offices in the 1850s. U.S. Census, 1860, 1880, McIntosh County.

11. Emma Carpenter Churchill (1819-1889), born Banburg, England, wife of William W. Churchill (1811-1888), born Oxfordshire, England. W. W. Churchill was a Darien store owner specializing in general merchandise who began his business in the town prior to the Civil War. Both are buried in St. Andrew's Cemetery near Darien. W. W. Churchill, for many years, was in business with William A. Wilcox (note 197) in Darien. U.S. Census, 1860, 1870, 1880, McIntosh County; Walter Howard, "Howard's Directory of Brunswick, St. Simons, Darien and St. Marys, Georgia for 1892" (Brunswick, Ga., 1892), hereinafter cited as Howard's Directory, 1892; White, Index to McIntosh County Cemetery Records, 12; A. C. McKinley's Sapelo Island Journal, 1869-76, 1886, see Robert L. Humphries, ed., *The Journal of Archibald C. McKinley* (Athens: University of Georgia Press, 1991).

12. Dr. R. B. Harris practiced medicine in Darien in the 1870s and 1880s and became a close friend of John G. Legare. *Darien Timber Gazette,* March 9, 1877; U.S. Census, 1880, McIntosh County.

13. Rice field dikes with tidegates to regulate the amount of river water admitted to irrigate the rice fields.

14. Henry F. Hoyt was pastor of the Darien Presbyterian Church in 1878-79. Sessional records, Darien Presbyterian Church, 1878.

15. Theodore Pitkin Pease (1813-1878) migrated to McIntosh County, Georgia, from Connecticut ca. 1839-40, becoming a successful planter and businessman. In 1847 he married Augusta Powell (1827-1909), daughter of Allen Beverly Powell of McIntosh County. They resided at the Thicket northeast of Darien. Pease was active in the naval stores industry, was a commissioner of the McIntosh County Academy (1848-1861 and 1866-1875), Mayor of Darien and Chairman of the McIntosh County

Commissioners in 1874-75, and an elder of the Darien Presbyterian Church, 1870-78. He owned considerable property in McIntosh County at the time of his death July 19, 1878 of consumption. Both Pease and his wife are buried in Darien's St. Andrew's Cemetery. Sullivan, *Early Days on the Georgia Tidewater*, numerous references; White, Index to McIntosh County Cemetery Records, 34; *Darien Timber Gazette*, May 16, 1874, June 18, 1875; July 26, 1878; U.S. Census, 1840, 1850, 1860, 1870, McIntosh County; Sessional records, Darien Presbyterian Church, 1870-1878; Buddy Sullivan and William G. Haynes, *History of the First Presbyterian Church of Darien,, 1736-1986* (Darien, Ga.: First Presbyterian Church, 1986), 21-24.

16. Reuben King Walker (1842-1915), Darien businessman and civic leader, was elected elder of the Darien Presbyterian Church in 1874. He was the son of James Walker (1816-1858) and Sarah Amanda (King) Walker (1817-1876), both of McIntosh County. In the late 1880s, R. K. Walker played a leading role in the development of the railroad in McIntosh County. A true capitalist in the mold cast by Henry Grady's vision of the New South, Walker was president of the Darien Shortline Railroad and allied himself with local timber interests. His vision was to expand McIntosh County's timber interests at the deepwater anchorage afforded by Sapelo Sound with the shuttling of timber by rail from the interior, efforts in which he was only partly successful. Walker was also proprietor of the Darien Telegraph Company. He had two brothers who were also active in the civic and business affairs of Darien and McIntosh County, James Walker, Jr. (note 25) and Joseph A. Walker (note 98). Amanda King Walker Papers, Collection 828, Georgia Historical Society, Savannah; Sullivan, *Early Days on the Georgia Tidewater*, numerous references to Walker and railroading; *Darien Timber Gazette*, March 1, 1890; Howard's Directory, 1892; White, Index to McIntosh County Cemetery Records, 45; Sessional records, Darien Presbyterian Church, 1874.

17. T. Shepherd Quarterman was elected elder of the Darien Presbyterian Church in 1874. He was Inspector of Timber and Lumber for Darien in 1882. Sessional records, Darien Presbyterian Church, 1874; *Darien Timber Gazette*, May 26, 1882.

18. Lucien B. Davis was elected elder of the Darien Presbyterian Church in 1875. He was Deputy U.S. Marshal in McIntosh County in 1876. Sessional records, Darien Presbyterian Church, 1875; *Darien Timber Gazette*, September 29, 1876.

19. Robert Lachlison, Darien lumberman. He came to McIntosh County with his brother James ca. 1852 and owned and operated the America Mills sawmill on Pumpkin Hammock (Union Island) near Darien after the Civil War. Robert and James Lachlison were related to the Hilton lumbering family of Darien (see also notes 28 and 30). Hilton Papers, Collection 387, Georgia Historical Society; Box 1, Folder 5 containing Hilton, *High Water*; and Folder 8 containing Lillian Fox Sinclair, "My Recollections of Darien in the Late Seventies and Eighties" (typed manuscript, n.d., cited hereinafter as Sinclair, Darien Recollections).

20. C. M. Quarterman, Darien businessman. He owned wharfage on the Darien River in the 1880s.

21. Archibald S. Barnwell (1833-1917), rice planter, was McIntosh County Representative to the Georgia Legislature in 1875. Barnwell leased Altamaha delta land and cultivated rice on Generals and Champneys islands during the postbellum period. Legare had a business relationship with A. S. Barnwell for many years. U.S. Census,

1880, McIntosh County; A. C. McKinley Journal; Sullivan, *Early Days on the Georgia Tidewater*, 201, 336, 472; *Darien Timber Gazette*, June 18, 1875; Sinclair, Darien Recollections; Howard's Directory, 1892.

22. Albert B. Curry was Supply Pastor for the Darien Presbyterian Church in 1876-77 and served as Pastor from 1879-1883. Sullivan and Haynes, *History of the First Presbyterian Church of Darien*, 25; Sessional records, Darien Presbyterian Church, 1876, 1879.

23. Benjamin W. Key served as pastor of the Darien Methodist Church in 1878-79. *Darien Timber Gazette*, August 16, 1878.

24. Samuel J. Pinkerton (1820-1906) was rector of St. Andrew's Episcopal Church in Darien on two occasions, 1856-1865 and 1879-80. He spent his youth in Virginia and Kentucky, was ordained to the ministry of the Episcopal Church in 1855 and served parishes in Brunswick, Georgia and Marietta, Georgia, following his service in Darien. Walter Birt Sams, *A History of Saint Andrew's Episcopal Church* (Darien, Ga.: St. Andrew's Episcopal Church, 1993), 11-12.

25. James Walker (1845-1926), Darien merchant and businessman. Walker served as both Mayor of Darien and McIntosh County Commission Chairman from 1876 to 1882 and again from 1888 to 1892. James Walker's younger brother was Joseph A. Walker (note 98) who, in 1890, was a McIntosh County commissioner and Darien city alderman at the same time James Walker was commission chairman and mayor. Also at this time, James Walker was an elder of the Darien Presbyterian Church and Joseph A. Walker was a deacon of the same church. In the 1880s and 1890s, James and Joseph Walker operated a dry goods and grocery business in Darien, the firm of J. and J. A. Walker. James and Joseph A. Walker were the sons of James Walker, Sr. and Sarah Amanda (King) Walker, both of McIntosh County. Their older brother, Reuben King Walker (note 16), was also active in the civic and business affairs of McIntosh County. *Darien Timber Gazette*, September 29, 1876, January 26, 1889, July 26, 1890, November 7, 1891; Amanda King Walker Papers, Collection 828, Georgia Historical Society; Sessional records, Darien Presbyterian Church, 1890; Howard's Directory, 1892.

26. Charles H. Hopkins, Jr. (1843-1899) served as Darien city marshal for eleven years in the late 1870s and the 1880s. C. H. Hopkins is listed as ordinary (probate judge) of McIntosh County in 1878, a position he held for ten years, and he was elected to the state legislature in 1892. His father was Charles Horrie Hopkins, Sr. (1812-1886), a leading McIntosh County planter, politician and lawyer in the late antebellum period, and owner of 106 slaves at his Belleville plantation. Legare relates in his journal of an unfortunate railroad accident which befell C. H. Hopkins, Jr. on March 3, 1897. *Memoirs of Georgia Containing Historical Accounts of the State's Civil, Military, Industrial and Professional Interests* (Atlanta: Southern Historical Association, 1895), 499; *Darien Timber Gazette*, August 23, 1878, April 4, 1887; *Eighth Census of the United States, 1860*, Slave Schedules; U.S. Census, 1880, McIntosh County; Howard's Directory, 1892; A. C. McKinley Journal; Sullivan, *Early Days on the Georgia Tidewater*, 284.

27. Alonzo Guyton was deputy sheriff of McIntosh County in 1874, and was one of several African-Americans holding local governmental positions locally during the Reconstruction period. Others during the period were James R. Bennett, McIntosh

County sheriff, and Lewis Jackson, county ordinary. From the late 1870s until 1902, Alonzo Guyton was deputy marshal for the City of Darien. See also note 170. *Darien Timber Gazette*, May 16, 1874, May 23, 1882; *Darien Gazette*, August 23, 1902; Russell Duncan, *Freedom's Shore: Tunis Campbell and the Georgia Freedmen* (Athens: University of Georgia Press, 1986), 77, 98-99.

28. Joseph Hilton (1842-1920), Darien lumberman and businessman, one of the most prominent citizens of McIntosh County ca. 1880-1910 during the postbellum lumber boom. Son of Thomas Hilton who migrated to McIntosh County with the Lachlison brothers, Robert and James, prior to the Civil War, Joseph Hilton saw war service then returned with his cousin, James Lachlison (note 30), to develop sawmills and other lumbering interests around Darien. He was president of the Hilton Timber & Lumber Company as that firm rose to prominence on the local scene with two Darien sawmills, Lower Bluff and Upper Mill, and a large mill at Doboy Island. The firm eventually had large mills in Savannah, Belfast (Bryan County) and Camden County on the Satilla River. Hilton organized the merger in 1888 with the Dodge interests on St. Simons Island to form the Hilton & Dodge Lumber Company, which, with headquarters in Darien and corporate offices in New York City, monopolized the Atlantic coast timber market for two decades. In 1892, Hilton & Dodge was listed with capital of $1 million. Hilton and his wife, Ida (Naylor) Hilton, were civic-minded and socially-active. They resided on Vernon Square in Darien, a home which still stands at this writing. He is buried in Bonaventure Cemetery in Savannah. Sullivan, *Early Days on the Georgia Tidewater*, 440-47, et al; Hilton, *High Water on the Bar*, 3-8, in Hilton Papers, No. 387, 1:5, Georgia Historical Society; Agreement for sale of St. Simons Mills to Hilton- Dodge Lumber Company, December 1, 1888 in Hilton Papers, No. 387, 2:12, GHS; Journal of Joseph Hilton, October 31, 1891 to November 30, 1893 for Hilton-Dodge Lumber Company in Hilton Papers, No. 387, 2:12, GHS; Abbie Fuller Graham, *Old Mill Days, 1874-1908* (Brunswick, Ga.: St. Simons Public Library, 1976); *Darien Timber Gazette*, October 27, 1888; U.S. Census, 1880, 1900, McIntosh County; Sinclair, Darien Recollections; Howard's Directory, 1892.

29. Isaac Means Aiken (born 1831), Darien lumberman, migrated to McIntosh County from South Carolina in 1854 and developed the Palmetto Mills sawmill on Hird Island prior to the Civil War. Hird Island was a marsh island between the Ridge on the mainland and Doboy Island on Doboy Sound (see also note 247). Aiken's mill had deepwater frontage on the North River. He lived on Hird Island, ran his sawmill during the Reconstruction period and became a leading citizen of McIntosh County. Aiken was Clerk of McIntosh County Superior in 1874-76. He built his Darien residence (Open Gates) on Vernon Square in 1876. Aiken's descendants became prominent in the business and civic affairs of Brunswick, Georgia. U.S. Census, 1860, 1870, 1880, McIntosh County; *Darien Timber Gazette*, June 18, 1875; A. C. McKinley Journal; Sullivan, *Early Days on the Georgia Tidewater*, 311, 382, 466 et al; Edwin H. Ginn, *The First Hundred Years* (Brunswick, Ga.: The American National Bank, 1989), 14.

30. James Lachlison (1837-1920), Darien lumberman and civic leader during the postbellum period. Lachlison's father, Robert, and uncle, James Lachlison, came to Darien with the Hiltons (notes 19, 28) ca. 1852 to begin their lumbering activities. In

the 1890s, he was superintendent of mills for the Hilton-Dodge Lumber Company. He served as a McIntosh County commissioner and Darien city alderman from the 1870s to the 1890s. His wife was Sarah Lachlison (1841-1904). They are buried in St. Andrew's Cemetery. Hilton, *High Water on the Bar*, 3-8; *Darien Timber Gazette*, May 16, 1874, May 21, 1891; White, Index to McIntosh County Cemetery Records, 26; U.S. Census, 1870, McIntosh County; Howard's Directory, 1892; Sinclair, Darien Recollections.

31. William Henry Atwood (1836-1912), McIntosh County businessman and politician during the postbellum period, son of Henry Skilton Atwood and Ann Margaret McIntosh. William H. Atwood served as a 1st lieutenant, then captain, with the McIntosh County unit of the Fifth Georgia Regiment during the Civil War. In 1871 he married Tallulah E. Butts (1850-1909) of Macon. Atwood resided at Cedar Point on the McIntosh County tidewater north of Darien, was a large landowner in the county, including holdings at the Sapelo Sound timber loading grounds and much of nearby Creighton Island, and was one of McIntosh's most prominent citizens for many years. He represented the county in the Georgia General Assembly in 1876-77 and was elected State Senator of coastal Georgia's 2nd District for the 1886-87 term. *Memoirs of Georgia*, 497-99; *Tax Digest*, McIntosh County, 1837; U.S. Census, 1860, 1870, 1880, 1900, McIntosh County; Sullivan, *Early Days on the Georgia Tidewater*, 274, 294, 686-88, et al; White, Index to McIntosh County Cemetery Records, 7.

32. Thomas Hart Gignilliat (1842-1905), McIntosh County rice planter, son of William Robert Gignilliat, Sr. and Helen Mary Hart. T. H. Gignilliat came from a family of antebellum rice planters in McIntosh County. He cultivated acreage on the Greenwood and Windy Hill rice tracts on Cathead Creek several miles west of Darien from the early 1870s until his death. His brother, W. Robert Gignilliat, Jr., was a Darien attorney. T. H. Gignilliat was a McIntosh County commissioner and Darien city alderman throughout the 1870s and 1880s. He is buried in St. Andrew's Cemetery. *Darien Timber Gazette*, May 16, 1874, July 29, 1888, September 26, 1890; White, Index to McIntosh County Cemetery Records, 21; Robert G. Kenan, *History of the Gignilliat Family* (Easley, S.C.: Southern Historical Press, 1977), 40.

33. Henry B. Tompkins was circuit judge for the Superior Court of McIntosh County in the mid-to-late 1870s. He was best remembered in Darien for his role in the Tunis G. Campbell, Sr., affair. Campbell (1812-1891) was an African-African former Freedmen's Bureau official who came to McIntosh County in 1867 from St. Catherines Island and established a powerful political base among the large number of blacks which outnumbered local whites by almost three-to-one (the U.S. Census of 1870 listed a population of 546 persons for Darien—435 black and 111 white—and 3,938 in McIntosh County outside of Darien—2,883 black and 1,085 white). Campbell represented McIntosh County in the Georgia Senate in the early 1870s. As Justice of the Peace in McIntosh County he wielded considerable influence and advanced the cause of local blacks. He was accused by the local press of conducting a "reign of terror" among local whites. Various abuses of power led to formal charges and indictments being issued against Campbell in 1875 with the cases being heard in Judge Tompkins' court. Campbell was sentenced to a year in the Georgia penitentiary by Tompkins in January 1876. He was freed in January 1877 but never regained his influence in

McIntosh County. Duncan, *Freedom's Shore: Tunis Campbell and the Georgia Freedmen*; Sullivan, *Early Days on the Georgia Tidewater*, 333-38, et al; A. C. McKinley Journal; U.S. Census, 1870, McIntosh County; *Darien Timber Gazette*, January 16, 1876, October 6, 1876.

34. Thomas Butler Blount (1842-1908), sheriff of McIntosh County, 1875 to 1891, and again from 1895 to 1903. A baker by profession, T. B. Blount was the son of Edmond M. Blount, Sr. of McIntosh County, a leading Darien civic leader prior to the Civil War. He and his wife, Mary J. Blount (1838-1919), are buried in St. Andrew's Cemetery. U.S. Census, 1880, 1900, McIntosh County; Howard's Directory, 1892; White, Index to McIntosh County Cemetery Records, 10.

35. James E. Holmes (1804-1883), known as "Dr. Bullie." Holmes, a native of Sunbury, Liberty County, Georgia, was the son of James and Mary (Kell) Holmes. He graduated from the University of Pennsylvania Medical School in 1825 and moved to Darien later that year where he remained for the rest of his life. In 1838 he married Susan Olivia Clapp and they resided at the Ridge near Darien for many years. Holmes served as plantation physician for many of the local rice plantations, including Butler's Island during which time he became a close friend of Pierce Mease Butler. He is mentioned by Fanny Kemble Butler in her *Journal* as "a shrewd, intelligent man, with an excellent knowledge of his profession, much kindness of heart and apparent cheerful good temper." For many years, Holmes served as health officer and port physician for the city of Darien, both before and after the Civil War. From 1875 to 1880 he wrote a popular column, seventy-six in all, of his antebellum reminiscences for editor Richard Grubb in the *Darien Timber Gazette*. He is buried in St. Andrew's Cemetery. Delma E. Presley, ed., *Dr. Bullie's Notes* (Atlanta: Cherokee Publishing Co., 1976); Frances Anne Kemble, *Journal of a Residence on a Georgian Plantation in 1838-1839*, edited by John A. Scott (New York: Alfred A. Knopf, 1961), 91, 92; Sullivan, *Early Days on the Georgia Tidewater*, numerous references; *Darien Timber Gazette*, 1875-1880; White, Index to McIntosh County Cemetery Records, 23.

36. William Robert Gignilliat, Jr. (1839-1885), Darien attorney and son of prominent antebellum planter W. R. Gignilliat, Sr. and Mary Hart Gignilliat. W. R. Gignilliat held several local administrative positions in the 1870s and 1880s. Kenan, *History of the Gignilliat Family*, 35; *Darien Timber Gazette*, July 21, 1882.

37. Louis Eugene Bree DeLorme (1830-1888), Darien attorney and son of Achilles A. and Mary (Lessare) DeLorme, the latter being postmaster of Darien in the 1850s and 1860s. L. E. B. DeLorme was admitted to the bar in 1859 and for many years was a prominent Darien attorney. He is buried in St. Andrew's Cemetery. White, Index to McIntosh County Cemetery Records, 16; Sessional records, Darien Presbyterian Church, 1870; *Darien Timber Gazette*, June 2, 1888.

38. Elihu S. Barclay (1832-1879), Darien inspector general of timber in 1874-75, Confederate veteran buried in St. Andrew's Cemetery. White, Index to McIntosh County Cemetery Records, 8; *Darien Timber Gazette*, May 16, 1874.

39. Steam rice mill engine (usually wood-burning) to operate machinery for threshing rice.

40. Doboy Island, on Doboy Sound south of Sapelo Island. Doboy was a major center for shipping coming into the local port to load timber and lumber, particularly

deeper-draft vessels unable to get to Darien itself ten miles west because of shallow water at low tide. A sawmill, general store, ship chandlery and stevedores' housing was located on the island. Another sawmill was just across the North River on Rock Island. Ships anchored in Doboy Sound or tied up at the Doboy wharves to load. The island was also a mail distribution center for the area and a main stop on the inland waterway steamboat route between Georgia and Florida. Sullivan, *Early Days on the Georgia Tidewater* is the most detailed source of material about Doboy; see also A. C. McKinley Journal; U.S. Coast Survey Topographical Map, "Doboy Sound and Vicinity, Georgia," 1868.

41. David Sutherland Sinclair (1852-1910), McIntosh County rice planter and close friend of J. G. Legare. D. S. Sinclair planted rice on Cathead Creek just west of Darien, at Sidon plantation, a tract of considerable acreage planted by James Smith, then the Dunwody family before the Civil War. He was an inspector of timber and lumber for Darien in the late 1880s and the early 1890s. Sinclair was also in the naval stores business and had turpentining activities at Sidon. A brother was Benjamin Thomas Sinclair (1861-1925), who ran a dry goods store on Broad Street in the 1890s. B. T. Sinclair was mayor of Darien in 1924-25. His wife was Lillian (Fox) Sinclair (1867-1949), a local educator who contributed much to the recording of Darien's early history. Another brother was William Waters Sinclair (1854-1912), city marshal of Darien in the early 1900s. Howard's Directory, 1892; White, Index to McIntosh County Cemetery Records, 39; *Darien Timber Gazette*, July 26, 1890; *Darien Gazette*, July 15, 1893; M. H. and D. B. Floyd Papers, No. 1308, 8:86, Georgia Historical Society.

42. The Ridge was a small community situated on high ground, a "ridge," overlooking Doboy Sound about three miles northeast of Darien. In the postbellum period, the Ridge was home to a number of bar pilots who guided timber shipping into Doboy Sound. The Ridge was also a summer residence for some of the wealthier Darien families, including the Hiltons, who wished to escape to a cooler locale, but still convenient to town. Families living in or near the rice delta customarily moved to the Ridge and other mainland areas during the summer to escape the so-called "miasma," or malarial airs, of the rice marshes. The Ridge became known officially as Ridgeville during the postbellum period when a post office was located there.

43. Jonesville was located in the northwest extremity of McIntosh County, about twenty-five miles by road from the county seat at Darien. During the antebellum plantation period, a number of families from nearby Liberty County had second homes at Jonesville. A church and school were located there. When the Florida Central & Peninsular Railroad passed through western McIntosh County in 1893, the focus of the community moved about a mile to the east adjacent to the tracks. Jonesville became known as Jones Station for several years, and later, simply as Jones. A small residential community now comprises present-day Jones.

44. Possibly the Blue and Hall Landing site. Suttons Landing is not shown on nineteenth century maps.

45. Ashantilly is a mile outside Darien off the Cow Horn (Ridge) Road. The residence called Ashantilly was built by Thomas Spalding of Sapelo in the 1820s for use as a mainland residence. The St. Andrew's Cemetery is also located at Ashantilly, adjacent to the Ashantilly house. During the postbellum period, a residential community

grew up at Ashantilly, which was located on the marshfront.

46. The Legare-Britt residence at Ashantilly is still standing. The Legare family had, for several years, resided in a frame dwelling on Generals Island. This structure, practically amid the rice fields themselves, was across the Darien River from the present Skipper Seafood docks. In 1888, Legare dismantled this dwelling and used many of the materials to build an addition to his Ashantilly residence.

47. Legare served as an elder and member of the Session for the Darien Presbyterian Church continuously from 1882 until his death in 1932, just short of fifty years.

48. The Shell Road connected Darien, Ashantilly and the Ridge. Known also as the Cow Horn Road, the road was paved with oyster shells during the 1890s. The route is the present day State Highway 99.

49. Possibly the Butler Cemetery just outside Darien.

50. Charles Rothschild, born in Germany in 1842, was a Darien merchant from the late 1870s to the 1890s. U.S. Census, 1880, McIntosh County; Howard's Directory, 1892.

51. Charles Oliver Fulton (1848-1911), Darien butcher and grocer. C. O. Fulton was one of the most enterprising merchants in McIntosh County during the Reconstruction and postbellum periods. He owned considerable acreage in the northwestern section of McIntosh County at Jonesville on which he raised cattle to supply his meat business. In 1873, Fulton opened the Darien Market, in which he specialized in meats and vegetables, and operated the business until his death in 1911. Shortly before his death Fulton purchased the Marsh Landing tract on Sapelo Island on which he also planned to raise cattle. He resided with his family on Black Island near Ashantilly. Sullivan, *Early Days on the Georgia Tidewater*, 818-19, et al; White, Index to McIntosh County Cemetery Records, 19; *Darien Timber Gazette*, April 3, 1875; Howard's Directory, 1892.

52. Thomas Bourke Spalding (1851-1884) of Sapelo Island was the son of Randolph and Mary (Bass) Spalding and a grandson of Thomas Spalding of Sapelo (1774-1851). In 1874 he married Patience (Ella) Barrow and lived on Sapelo Island at Marsh Landing until his untimely death in 1884 from a hunting accident. Bourke Spalding, his brother Thomas Spalding (1847-1885) and brother-in-law Archibald C. McKinley (1842-1917) all lived on Sapelo Island during the Reconstruction period and raised cattle for the purpose of selling beef to the shipping in Sapelo and Doboy sounds during the winter and spring timber season at Darien. They also operated a small steamboat which transported passengers and freight from Doboy Island to Darien. Further tragedy struck this family only a few months later when, in January 1885, Thomas Spalding was accidentally killed in a railroad accident in Macon, Georgia. Sullivan, *Early Days on the Georgia Tidewater*, 368-409, is the most detailed account of the Spalding family and this period of Sapelo Island's history; A. C. McKinley Journal; Spalding Family Papers, Collection 750, Georgia Historical Society, Savannah. White, Index to McIntosh County Cemetery Records, 40.

53. St. Andrew's Cemetery, adjacent to the Ashantilly community overlooking the marshes and Black Island, was originally the Spalding family burial ground. Many members of the Spalding and allied families are buried there. In 1867, the last surviv-

ing son of Thomas Spalding of Sapelo, Charles Harris Spalding (1808-1887), deeded St. Andrew's for use as a public cemetery.

54. Richard Lewis Morris (1818-1885), Darien rice planter and businessman. R. L. Morris and his brother, Charles M. Morris, cultivated rice on the Cathead Creek tracts of Ceylon and Potosi west of Darien before the Civil War. R. L. Morris served as a McIntosh County commissioner and Darien alderman during the 1870s. Sullivan, *Early Days on the Georgia Tidewater*, 224-27, et al; A. C. McKinley Journal; *Darien Timber Gazette*, May 16, 1874.

55. Inwood was north of the Ridge on the Cow Horn Road, about five miles from Darien.

56. Rhett's Island was immediately east of Generals Island. On most maps of the nineteenth century Rhett's is shown as a part of Generals Island. Generals Cut, a narrow waterway which connected Darien to the Butler River, separated Generals and Rhett's, which was named for Robert Barnwell Rhett of Charleston, South Carolina who grew rice there in the 1850s. U.S. Coast Survey Topographical Map, Altamaha Sound and Vicinity, Georgia, 1869.

57. Henry Todd (1816-1886), successful black Darien businessman. Before the Civil War, Henry Todd was known as a "free man of color." After the war, he owned and operated the Sansavilla Steam Saw Mill in Darien, had a considerable amount of real estate in his possession and amassed some amount of wealth. He and his wife were members of the white Presbyterian Church and were highly-respected members of the Darien community. At his death, Todd left endowments for black and white churches and schools in Darien. The Todd School in Darien educated several generations of black children. He and his wife Mary Ann Cardone (1826-1887) are buried in Darien's Upper Mill Cemetery. Sinclair, Darien Recollections; McIntosh County Probate Court, Will of Henry Todd, dated April 12, 1882; *Darien Timber Gazette*, May 8, 1886.

58. Upper Mill Cemetery on River Road in Darien, formerly known as the Presbyterian Cemetery. Sessional records, Darien Presbyterian Church.

59. The Mallard family had cultivated rice on Cathead Creek at Oasis plantation since the antebellum period. This was begun in the 1850s by Thomas S. Mallard of Liberty County (see note 150).

60. William Nightingale was the son of Phineas Miller Nightingale, a McIntosh County rice planter before the Civil War. The elder Nightingale inherited Dungeness plantation on Cumberland Island and planted Sea Island cotton there until the 1850s. William Nightingale planted rice at Cambers Island (note 80) in the Altamaha River delta for several years. Thomas Porcher Ravenel Papers, Collection 649, Box 20, Folders 64, 65, Georgia Historical Society, Savannah.

61. Mayhall (pronounced "my-hall") Island was east of Darien on the Darien River near Black Island. A sawmill operated there before and after the Civil War. U.S. Geological Survey map, Darien Quadrangle, 1918.

62. For another perspective on this event see the comments of Archibald C. McKinley of Sapelo Island contained in Sullivan, *Early Days on the Georgia Tidewater*, 399; also *Darien Timber Gazette*, September 4, 1886.

63. For a detailed review of local steamboat activity during this period see Sullivan, *Early Days on the Georgia Tidewater*, 434-39, and the A. C. McKinley Journal.

64. Hammersmith Landing was on the Glynn County side of the Altamaha delta at Altama plantation, just east of Hopeton. A ferry ran between Darien and Hammersmith Landing, from which point travelers could make rail connections to the interior at Sterling.

65. James Troup Dent (1848-1913) owned and managed the Hofwyl-Broadfield rice plantation on the Glynn County side of the Altamaha delta across from Darien. This was one of the oldest continuously-run rice plantations in the delta, being owned by the same family since 1807 when William Brailsford (ca. 1760-1810) and his wife, Maria Heyward, of South Carolina acquired the tract. Their daughter, Camilla Brailsford, married James McGillivray Troup of Darien in 1814. Troup, at his death in 1849, had 7,300 acres of land and 357 slaves but part of his property was heavily mortgaged. The Troup's daughter, Ophelia Troup, married George Columbus Dent in 1847. It was Dent who provided the name of Hofwyl to the plantation, being the name of a boys school he attended in Switzerland. The present plantation house at Hofwyl was ostensibly built by the Dents in 1851 as an overseer's residence, but it ultimately became the family home for several generations before being acquired by the State of Georgia upon the death of the last surviving Dent, Ophelia Dent, in 1973. The James Troup Dent referred to by Legare in his Journal was the eldest son of George and Ophelia Dent. In 1880, J. T. Dent married Miriam Gratz Cohen, an energetic woman who helped bring the plantations back into Dent family ownership after both Hofwyl and Broadfield were foreclosed on after the Civil War. Rice was grown at Hofwyl until about 1915. Victoria Reeves Gunn, "Hofwyl Plantation," (Georgia Department of Natural Resources, unpub. ms., 1975), 72-79; Dent Family Papers, Collection 213, Georgia Historical Society, Savannah; Howard's Directory, 1892.

66. Sarah F. Bealer (1861-1886), first wife of Lewis Myers Bealer (1857-1942) of Darien. Both are buried in St. Andrew's Cemetery. White, Index to McIntosh County Cemetery Records, 9; Sessional records, Darien Presbyterian Church, 1903; Howard's Directory, 1892.

67. Legare was referring to the almost total destruction of Darien on June 11, 1863 by Union forces stationed on nearby St. Simons Island, the 54th Massachusetts Volunteer Regiment and the 2nd South Carolina Volunteers, two black units officered by whites. The commanding officer of the raid, Col. James Montgomery, issued the orders to burn the town. The 54th Massachusetts commander, Col. Robert Gould Shaw (whose parents were abolitionists as was their friend Frances Anne Kemble), reluctantly obeyed orders to sack Darien. He was killed in action leading his troops against Battery Wagner, South Carolina, about a month later. In 1870, Shaw's mother contributed funds to build an Episcopal church on the Ridge after the St. Andrew's Episcopal Church in Darien was burned during the Union raid. Darien was not burned by units of Sherman's army, a commonly held, but mistaken, assumption. Sherman's march to the sea occurred more than a year after the burning of Darien. The entire Darien waterfront and commercial district was destroyed in this controversial incident which created resentment on the part of residents who had evacuated the town. See the author's introduction to this volume; also, Sullivan, *Early Days on the Georgia Tidewater*, 292-309.

68. The Masonic Hall was located on the upper bluff of the waterfront just east

of the present bridge over the Darien River. Sanborn Insurance Map of Darien, 1885.

69. The Magnolia House hotel was located on the upper bluff on the east side of the present bridge over the Darien River. The hotel fronted on both Broad Street and the river, where it had a steamboat wharf for the convenience of arriving guests. Opened in the early 1870s, the Magnolia was the best hotel in Darien and was a favorite of travelers on the inland waterway steamboat route, and also for visitors to Darien on timber business and locals who enjoyed the Magnolia's restaurant. In 1874, manager A. E. Carr was advertising to "all who desire first class accommodations" room and board at $2 per day. The Magnolia was not rebuilt after the 1887 fire. *Darien Timber Gazette*, May 23, 1874; Sanborn Insurance Map of Darien, 1885; Fred Cook, principal investigator, with Charles Pearson, Buddy Sullivan, Elizabeth Misner and Elizabeth Reitz, "An Historical and Archaeological Study of the Darien, Georgia Waterfront," for McIntosh County Industrial Authority, 1991.

70. The office of editor Richard Grubb and his *Darien Timber Gazette* was located on the upper bluff near the hotel on Broad Street. Sanborn Insurance Map of Darien, 1885.

71. James Edward Holmes (1840-1890), inspector general of timber and lumber for Darien in the 1880s. He was the son of Dr. James Holmes of Darien (note 35). *Darien Timber Gazette*, September 29, 1888; White, Index to McIntosh County Cemetery Records, 22.

72. The tabby ruins of these buildings, which date to ca. 1819, are still standing on the Darien waterfront. See the photographs in Sullivan, *Early Days on the Georgia Tidewater*, 351-53.

73. Legare is in error here as he considerably overstates the distance, which did not exceed 250 river miles.

74. Legare's rice planting operations on Generals Island had the advantage of being one of the most easterly, and therefore further downstream, of the other Altamaha delta plantations. This resulted in his being less severely affected by the periodic freshets than those upriver.

75. Cathead Creek, on the west side of Darien, flowed from Buffalo Swamp into the Darien River. Since it experienced the same tidal fluctuation as the rest of the Altamaha, Cathead Creek had a number of rice growing operations before and after the Civil War.

76. The Legares returned to Darien on November 11th.

77. A ship in port to load lumber at one of the Darien sawmills.

78. Snow Creek, a tidal stream, flowed between Black Island and Ashantilly. Its southern end emptied into the Darien River at Lower Bluff. After ca. 1950 Snow Creek is shown on charts as Black Island Creek. U.S. Coast Survey Topographical Map, "Altamaha Sound and Vicinity, Georgia," 1869; U.S. Department of Agriculture, Soil Survey map, McIntosh County, 1929.

79. Hopeton plantation on the Glynn County side of the Altamaha delta, referred to as the "model plantation of the antebellum South" before the War Between the States when it was managed and partly owned by James Hamilton Couper. During the postbellum period, Richard Corbin, a descendant of James Hamilton, one of the co-owners of the tract, cultivated rice at Hopeton and lived at Altama on the adjoining

tract. James M. Clifton, "Hopeton: Model Plantation of the Antebellum South," *Georgia Historical Quarterly* 66 (Winter 1982).

80. Cambers Island, in McIntosh County across the river from Hopeton, was the Altamaha delta rice plantation of Thomas Spalding in the antebellum period until 1855 when it was acquired by P. M. Nightingale. The Nightingales grew rice there before and after the Civil War. Cambers Island was located just west of Butler's Island. Cambers and Hopeton were the westernmost of the Altamaha River rice tracts.

81. Rafts, or drifts, of timber logs lashed together at the Darien timber booms where timber was gathered for inspecting and grading by public, city-appointed inspectors, prior to sale to the local sawmills.

82. Octavius C. Hopkins, Jr. (1863-1909), inspector of timber and lumber in Darien from the 1880s until the early 1900s. He was the son of O. C. Hopkins, Sr. (1819-1881) of McIntosh County. White, Index to McIntosh County Cemetery Records, 24; *Darien Timber Gazette,* July 7, 1882, December 14, 1901.

83. At the Romerly Marsh passage which at that time constituted the inland waterway route just south of Savannah between Skidaway and Wassaw islands. U.S. Coast and Geodetic Survey, "Atlantic Local Coast Pilot," 1885 edition; U.S. Army Corps of Engineers, "Intracoastal Waterway, Beaufort, N.C. to Key West, Fla. Section," House of Representatives, 63rd U.S. Congress, Washington, D.C., 1913; A.C. McKinley Journal.

84. N. Keff Smith served on two occasions in the pulpit of the Darien Presbyterian Church, the first from 1889 to 1892, and the second from 1901 to 1908. He became a close friend of Legare and was a frequent guest of the Legares in the years following his pastorates in Darien. Sessional records, Darien Presbyterian Church, 1889, 1901; Howard's Directory, 1892.

85. In fact, Legare served as clerk of Session for Darien Presbyterian Church for thirty-nine years, until 1928.

86. Joseph Mansfield (1843-1912), born in Ireland, was a Darien businessman and merchant who came to McIntosh County in the late 1870s. He acquired property at the Thicket and resided there for a number of years. In the late 1880s and early 1890s he is listed as a member of the Darien Board of Health. In the 1890s, he represented McIntosh County in the Georgia General Assembly. U.S. Census, 1880, McIntosh County; *Darien Timber Gazette,* September 29, 1888; Howard's Directory, 1892.

87. Henry A. Weil and his brother, Simon A. Weil, were Darien merchants in the 1880s and 1890s. H. A. Weil ran a dry goods store, specializing in clothing for women and children. Simon Weil was a grocer. *Darien Timber Gazette,* April 30, 1887; Howard's Directory, 1892.

88. Eulonia was formerly known as Sapelo Bridge and was the first designated seat of government for McIntosh County when it was created from Liberty County in 1793. Sapelo Bridge was in the center of the county and at the head of navigation of the Sapelo River, a tidal stream emptying into Sapelo Sound. The name of Sapelo Bridge was officially changed to Eulonia in 1895 for purposes of eliminating confusion between the post office there and the one at Sapelo on Hazzard's Island, Front River, Sapelo Sound (east of Creighton Island). Eulonia was supposedly named for Eulonia, South Carolina by the postmaster, O. S. Davis. Sullivan, *Early Days on the Georgia*

Tidewater, 594-97.

89. Work on the Darien Short Line Railroad began in 1889 with plans for the road to run from Tattnall and Liberty counties to Belleville and Sapelo Sound in McIntosh. Its purpose was to transport timber from the interior to the deepwater anchorage near Creighton Island on Sapelo Sound. This was an investment venture by local businessmen, headed by Reuben K. Walker of McIntosh County (note 16), with the help of northern capital. The company eventually ran out of money before the completion of the project. The Short Line was acquired by the Darien and Western Railroad Company with track being extended to Darien in 1895. Sullivan, *Early Days on the Georgia Tidewater,* 499-505; *Darien Timber Gazette,* August 3, 1889.

90. Crescent was a community which developed in the 1890s as a result of the Darien Short Line, then Darien & Western railroads passing through. It acquired a post office and eventually became the site of an important maintenance and watering facility for the railroad. It was named for the bend, or "crescent," in the Sapelo River which flowed past Belleville to the north of Crescent.

91. Belleville was located just north of Crescent on the Sapelo River. Railroad investors had plans of developing Belleville as a port for the shipment of timber brought in by rail from the interior. Belleville was conveniently located to the deepwater anchorage of Sapelo Sound. It was formerly the site of Belleville plantation of Francis Hopkins and his sons during the antebellum period.

92. The Thicket was located six miles northeast of Darien off the Cow Horn Road fronting the marsh overlooking Doboy Sound and Sapelo Island. In antebellum times it was the site of a sugar mill and rum distillery operated by William Carnochan, with architectural design and capital provided by Thomas Spalding of Sapelo. This operation ended after the mill was severely damaged in the 1824 hurricane, but the tabby ruins still remain at the edge of Carnochan Creek. Later it was Spalding land where Sea Island cotton was cultivated, and the remains of several tabby slave cabins still stand at the site. The Thicket and the long neck of land running northeast, known locally as Pease Point, were owned by T. P. Pease after the Civil War. Sullivan, *Early Days on the Georgia Tidewater,* 101-07; Marmaduke Floyd, "Certain Tabby Ruins on the Georgia Coast," in *Georgia's Disputed Ruins* (Chapel Hill: University of North Carolina Press, 1937), 111-132; M. H. and D. B. Floyd Papers, Collection 1308, 14:127-133, Georgia Historical Society, Savannah.

93. William C. Wylly (1842-1923), McIntosh County rice planter and businessman. During the postbellum period he cultivated rice on Broughton Island (note 94). In the 1890s, through his marriage to the widow of Thomas Spalding (II), Wylly lived on the south end of Sapelo Island and had some activity in the commercial harvest of shellfish for a time. He served as tax collector of McIntosh County several times, being listed in that capacity in 1882 and 1912. White, Index to McIntosh County Cemetery Records, 47; *Darien Timber Gazette,* May 26, 1882; *Darien Gazette,* January 20, 1912; Howard's Directory, 1892; Sinclair, Darien Recollections.

94. Broughton Island, a McIntosh County tract on the south branch of the Altamaha River across from the Glynn County tract of Hofwyl-Broadfield, was the scene of rice cultivation starting in the late 18th century. The rice plantation on Broughton was one of the most active during the antebellum period, being planted by

James M. Troup and Thomas Bryan Forman. It was well-developed prior to the Civil War with canals and irrigation ditches, a rice mill and slave housing. William C. Wylly cultivated rice on Broughton Island in the 1890s. It was one of the more isolated of the Altamaha delta rice plantations. Ruins of the rice mill and the canal system still remain at Broughton. Sullivan, *Early Days on the Georgia Tidewater,* 172-75, 770-72, 799-801; Georgia Bryan Conrad, "Reminiscences of a Southern Woman" (Hampton, Virginia, 1901); U.S. Coast Survey Topographical Map, "Altamaha Sound and Vicinity, Georgia," 1869.

95. Egg Island is a small marsh island with beach and high ground located at the mouth of the Altamaha River where it meets the Atlantic Ocean—between Wolf Island and Little St. Simons Island.

96. John S. Stebbins (born 1848), son of Charles Austin Stebbins and Margaret M. Stebbins of Liberty County, Georgia. Charles Austin Stebbins acquired Marengo plantation near Harris Neck, McIntosh County, prior to the Civil War. The Stebbins family lived at Marengo in the years after the war. John S. Stebbins eventually moved to Darien. U.S. Census, 1870, McIntosh County; Sullivan, *Early Days on the Georgia Tidewater,* 239-40. Riceboro was a small community at the headwaters of the North Newport River in Liberty County, about six miles north of the McIntosh County line.

97. J. K. Clarke and Company was one of Darien's most active timber firms with capital amounting to $100,000 in 1892. James K. Clarke, president of the concern, was born in England in 1837, began his timber business in Darien in 1875 and was successful through the 1880s and 1890s. He was active in local politics and a member of the McIntosh County Democratic Committee. In the 1880s, James K. Clarke served as British Vice-Consul to Darien. During this period, Darien had five vice-consulates: Portuguese (James Hunter), Brazilian (James E. Holmes), Norwegian and Swedish (John J. Kirby), German (August Schmidt) and British (James K. Clarke). At the time of this entry in Legare's Journal, J. K. Clarke and Company had just expanded its operations to Sapelo Sound where it had built wharfage on the Front River near Creighton Island. J. J. Kirby was secretary-treasurer of the firm. Two other large Darien firms, Hilton-Dodge and Hunter-Benn, had also begun loading activities at Sapelo Sound to accommodate larger ocean-going vessels entering the harbor to load local timber (see note 157). *Darien Timber Gazette,* July 21, 1882 and various issues, 1875-1892. Sullivan, *Early Days on the Georgia Tidewater,* 488-98; Howard's Directory, 1892; Sinclair, Darien Recollections.

98. Joseph A. Walker (1853-1911), Darien businessman and civic leader. With his brother, James Walker (note 25), Joseph A. Walker was proprietor of one of Darien's leading emporiums. He served as a McIntosh County commissioner and Darien city alderman. Joseph A. Walker was mayor pro-tem of Darien in 1895. *Darien Timber Gazette,* November 7, 1891; Howard's Directory, 1892; Amanda King Walker Papers, Collection 828, Georgia Historical Society, Savannah.

99. The Rev. John Ward Motte was a first cousin of Legare and had served as Rector of St. Andrew's Episcopal Church in Darien since June 1889.

100. Spalding Kenan (1836-1908), local physician and mayor of Darien from 1892-1901. Spalding Kenan was the son of Michael J. Kenan and Catherine Spalding Kenan of Milledgeville, Georgia and Sapelo Island, and grandson of Thomas Spalding

of Sapelo. He inherited from his parents a large tract of land on Sapelo Island, formerly a Sea Island cotton plantation, and lived there until ca. 1880, after which he removed to Darien to establish his medical practice. He was mayor of Darien during the worst racial unrest in the town's history in August-September 1899, which Legare records in detail in his Journal (see the Introduction to this volume). Sullivan, *Early Days on the Georgia Tidewater*, numerous references; A. C. McKinley Journal; Howard's Directory, 1892; White, Index to McIntosh County Cemetery Records, 25; *Darien Gazette*, March 31, 1900.

101. John Michael Fisher (1833-1913), Darien merchant and store owner, was a resident of Ashantilly and neighbor and friend of Legare. He was born in Germany, emigrated to Savannah in 1859 and arrived in Darien in 1865. In ca. 1870, John M. Fisher built a general store, saloon and bakery on the northwest corner of Broad and Screven streets in Darien. He was a baker of great popularity, as he baked fresh bread and rolls daily in his bakery which he built with a brick oven and chimney. The adjoining property was occupied for many years by the jewelry business of William McWhir Young. J. M. Fisher and his wife, Christina Leontina Fisher (1841-1889), lived in Darien east of the courthouse in a home they bought and later sold to local timber dealer August Schmidt. This house still stands and is known locally as the Schmidt house. The Fishers later built a home at Ashantilly. Fisher was a McIntosh County commissioner and Darien city alderman in the 1880s. He is buried in St. Andrew's Cemetery. Interviews with Annie Fisher Gill, December 29, 1996, January 8, 1997; U.S. Census, 1870, 1880, McIntosh County; *Darien Timber Gazette*, May 26, 1882; Duncan, *Freedom's Shore*, 90, 91, 96, 99, 103, 115; Howard's Directory, 1892; Sanborn Insurance Map of Darien, 1878, 1885.

102. Richard W. Grubb (1852-1918) was the editor and owner of the local weekly newspaper, the *Darien Timber Gazette*, which he established in the spring of 1874 (the *Timber* was dropped from the newspaper's name in 1893). Grubb developed a statewide reputation for his humorous, witty writing style. Grubb was never known to pull his editorial punches, and because of this his paper became one of the most popular in the state. Grubb diligently reported the activities of the local timber markets, noted the changing fluctuations in the prices of timber and tracked the arrivals and departures of the shipping in the port to load timber and lumber. Grubb and Legare were the same age and were close friends as they came to Darien within three years of each other and stayed for the remainder of their lives. It is the *Darien Timber Gazette* which provides a wealth of resource material for McIntosh County history during the Reconstruction and postbellum period of Grubb's editorship. See the author's Introduction to this volume for additional material. Sullivan, *Early Days on the Georgia Tidewater*, 353-57, et al; Presley, ed., *Dr. Bullie's Notes*, xx-xxv; White, Index to McIntosh County Cemetery Records, 21; Howard's Directory, 1892.

103. Charles Louis Livingston (1846-1901), Darien attorney and businessman. He was the son of Louis and Elizabeth Bass Livingston of Columbus, Georgia, and a cousin of Thomas Bourke Spalding and Thomas Spalding (II) of Sapelo Island. He was a resident of Sapelo Island before establishing his law practice in Darien in the mid-1880s. White, Index to McIntosh County Cemetery Records, 27; A. C. McKinley Journal; *Darien Timber Gazette*, September 29, 1888; Howard's Directory, 1892.

104. Wyatt deReviero Barclay (1861-1923), Darien attorney. He was the son of E. S. and Helen Stanford Barclay. W. deR. Barclay was practicing law in Darien as early as 1884 and was serving as clerk of McIntosh County Superior Court in 1885. He served as mayor of Darien from 1918 to 1923 and was active in local civic and community affairs throughout his career. *Darien Timber Gazette*, September 5, 1884, January 31, 1885; Howard's Directory, 1892; White, Index to McIntosh County Cemetery Records, 9.

105. William Konetzko (1843-1897) was a merchant and operated a saloon in Darien. He is buried in St. Andrew's Cemetery. His son, Arthur Konetzko (1875-1930), ran his father's store in Darien until ca. 1902. William Konetzko had a country home just south of the Ridge, which had previously been owned by Capt. A. S. Barnwell. Howard's Directory, 1892; White, Index to McIntosh County Cemetery Records, 26.

106. Bateau—a small wooden boat with raked bow and stern and flaring sides, usually flat-bottomed, but also built with the traditional V-hull. These were the workboats of the Georgia tidewater. They proliferated the coast from ca. 1890 to ca. 1960, being particularly useful in connection with the thriving Georgia oyster industry during that period.

107. Champneys Island was named in the early 1800s for John Champneys, a contemporary of Major Pierce Butler who owned neighboring Butler's Island in the Altamaha delta. Several planters successfully cultivated rice on Champneys during the antebellum period, including John Champneys Tunno, Hugh Fraser Grant, owner of Elizafield plantation across the river on the Glynn County side, and Jacob Barrett of South Carolina who leased, then owned, Champneys Island. The island is referred to as Tunno's Island in the Fanny Kemble Butler Journal and is shown on some late antebellum maps as Barrett's Island. After the Civil War, Champneys was owned by A. S. Barnwell for several years. See the Introduction to this volume for chain of title; also, Sullivan, *Early Days on the Georgia Tidewater*, 199-202, et al; Kemble, *Journal of a Residence on a Georgian Plantation*, 154-55; U.S. Coast Survey Topographical Map, "Altamaha Sound and Vicinity, Georgia," 1869.

108. The Victorian-style Oglethorpe Hotel opened in January 1888 on Newcastle Street in Brunswick, convenient to the local waterfront. It immediately became the premier hostelry on the coast between Savannah and Jacksonville, serving arriving members of the Jekyll Island Club, timber and naval stores dealers and many others. It was torn down in 1958.

109. Grant Chapel continues to play an important role in the religious life of Darien.

110. On the Darien River waterfront at the site of the present Skipper family shrimp docks.

111. Present site of the Darien News building adjacent to the bridge.

112. Ann Susan (King) (Anderson) Muller (1818-1892) of Harris Neck, McIntosh County. She was the daughter of antebellum cotton planter William John King of Harris Neck. John Muller (1830-1908) was the King's Swiss overseer for the plantation when he married Mrs. Anderson, then a widow. They continued to live at Harris Neck after the Civil War. John Muller was a McIntosh County commissioner in the 1890s.

Sullivan, *Early Days on the Georgia Tidewater*, 249-50, et al; Isabel Thorpe Mealing, *History of Northern McIntosh County* (Darien, Ga.: privately printed, 1992), 9; *Darien Gazette*, August 17, 1895.

113. Harris Neck is in the northeastern section of McIntosh County, its upper end fronting the South Newport River on the boundary with Liberty County and its south end accessible to the Julianton River and Sapelo Sound. The north and central sections of the Neck were the scene of extensive plantation activities during the period 1790-1861 by several different families, including Harris, Thomas, Gould, King and Delegal, with the Levett-Bennett family on the south end. Of particular interest to McIntosh Countians during the postbellum period was the purchase in 1889-90 of a tract of former Thomas land on the north end of Harris Neck by Pierre Lorillard, the tobacco magnate from New York City. Lorillard built a home and other structures. Foundation ruins of these still overlook the South Newport River. The history of Harris Neck is as intriguing as any area of tidewater Georgia. The most complete account is contained in Sullivan, *Early Days on the Georgia Tidewater*, 227-271, 498-501, 745-46, et al., see index; for Lorillard, see *Darien Timber Gazette*, February 15, 1890.

114. The Wolf Island Club was established in 1890 on five acres on the south end of Wolf Island, McIntosh County, which lay between Sapelo Island and Little St. Simons Island. Wolf Island was almost totally comprised of tidal salt marsh, except for a strip of beach along the ocean side of the island. There was a range beacon on the north end of the island where a keeper and his family lived. The Wolf Island Club was formed by businessmen from Darien and other parts of the state for recreational pursuits. Sullivan, *Early Days on the Georgia Tidewater*, 505-06.

115. Henry Kollock Rees (1822-1893), Rector of St. Andrew's Episcopal Church, was the son of Ebenezer Senior Rees (1790-1842) and Mary Dews Rees (1802-1853). E. S. Rees was cashier of the Bank of Darien when it was one of the most influential financial institutions in Georgia during Darien's heyday as a major shipping center for inland-grown cotton in the 1820s and 1830s. Henry K. Rees was first a Presbyterian, then Episcopalian, clergyman. He is buried in St. Andrew's Cemetery. White, Index to McIntosh County Cemetery Records, 36.

116. Frances Butler Leigh (1838-1910) was the daughter of Pierce Mease Butler (1810-1867) of Philadelphia and Frances Anne Kemble Butler (1809-1893), the well-known English actress. In a family torn by the issue of slavery and southern state's rights, Frances Butler (Fan), along with her father, sympathized with the South, while her mother and older sister, Sarah Butler Wister (1835-1908), had strong abolitionist sentiments. Following the Civil War, Frances Butler and her father renewed the family's rice growing operation on Butler's Island. In 1871 she married the Rev. James Wentworth Leigh, an English rector, and the two managed the Butler's Island plantation for several years before moving permanently to England. Mrs. Leigh leased her Butler's Island lands for rice cultivation to several local planters before finally selling the island in 1906. Sarah Butler Wister's son, Owen (Dan) Wister (1860-1938), eventually sold most of the Butler family's Georgia lands in the early 1900s. He is perhaps best known for writing the novel *The Virginian*, in 1902. Malcolm Bell, *Major Butler's Legacy: Five Generations of a Slaveholding Family* (Athens: University of Georgia Press,

1987); Sullivan, *Early Days on the Georgia Tidewater*, 338-44, et al; Sinclair, Darien Recollections; Thomas Porcher Ravenel Papers, No. 649, 20:60 containing Leigh-Wister papers, 1889, Georgia Historical Society, Savannah.

117. The 1893 yellow fever epidemic in Brunswick claimed fifty-three lives. Yellow fever was a recurring problem in port cities of the south Atlantic coast in the nineteenth century. Usually fatal with a short incubation period lasting only about a week, the disease was commonly introduced from merchant vessels arriving from the tropics during the warm weather months. Yellow fever epidemics afflicted Darien and Savannah in 1854. The worst epidemic was that of 1876, which took 1,066 lives in Savannah and 112 in Brunswick. In response, the U.S. Marine Hospital Service established the South Atlantic Quarantine Station at Blackbeard Island, McIntosh County, in 1880. Ships bound for ports in Georgia and South Carolina, including the heavy amount of timber shipping arriving in Darien waters, were required to be inspected for yellow fever at Blackbeard Island. A head surgeon and a six-to-seven person support staff operated the facilities at Blackbeard, which included a hospital, dormitory, disinfection wharf and various other facilities. A documented account of the Blackbeard quarantine station, which was deactivated in 1909, is contained in Sullivan, *Early Days on the Georgia Tidewater*, 474-88, et al.

118. The Darien Bank was founded in 1889. Up to that time, Darien had not had a bank since the Bank of Darien lost its charter in 1842. The Darien Bank was established in deference to the local timber business which was reaching its peak. Darien was then a prosperous community and the bank was able to begin on a sound footing. The first president of the bank was local business leader Adam Strain (1889-97), followed by timber merchant August Schmidt (to 1901), merchant William II. Strain (to 1906), R. H. Knox of the Hilton-Dodge Lumber Co. (to 1912), Elisha M. Thorpe, president of the McIntosh Naval Stores Co. (to 1915), and businessman C. M. Tyson (to 1921). Sullivan, *The Darien Bank: A Celebration of 100 Years*, 24-40; Howard's Directory, 1892.

119. The hurricane of 1893 was one of the worst to ever strike the Georgia-South Carolina coast. It dealt a severe blow to tidewater rice growing operations, particularly in South Carolina.

120. Sidon plantation was on Cathead Creek just west of Darien. Rice had been cultivated there since the early 1800s by the Smith and Dunwody families, and later by David S. Sinclair (note 41), who also had turpentining activities on the site. Archaeological investigations conducted at the Sidon site in 1993 revealed considerable evidence of rice growing, including foundation remains of tabby slave houses and remnants of farm implements spanning a period from ca. 1820 to ca. 1900. Sullivan, *Early Days on the Georgia Tidewater*, 223-25, 826-27.

121. The site of Barrington Station was about ten miles west of Darien, just north of the present Cox community. Nothing remains of Barrington. The last train ran through McIntosh county in 1985 when the tracks of the former Seaboard Air Line R.R. were taken up. Sullivan, *Early Days on the Georgia Tidewater*, for Barrington, 574-76, for the Florida Central & Peninsular Railroad, 504-05.

122. Altama was once part of Hopeton plantation on the Glynn County side of the Altamaha River. James Hamilton Couper, half owner and manager of Hopeton through much of the antebellum period with extensive rice and sugar operations,

acquired Altama as his own property and built the residence there in 1857. Previously, this site was the Woodville tract owned by Pierce Butler, where Butler slaves too old or infirm to work in the rice fields were housed. Hammersmith Landing was on the Woodville-Altama tract. Sullivan, *Early Days on the Georgia Tidewater*, 774-75, 808.

123. James S. Townsend (1840-1918), son of William Riley Townsend (1806-1874) and Dorinda Townsend, both of McIntosh County. Laurel Grove (also Kell's Grove) on Snow Creek near Lower Bluff (Fort King George site) was the home of the Kell family during the antebellum period. Settled by John and Margery Spalding (Baillie) Kell prior to 1820, Laurel Grove was the ancestral home of their son, John McIntosh Kell (1823-1900). The latter was the executive officer of the CSS *Alabama*, the most successful Confederate raider of the Civil War. Norman C. Delaney, *John McIntosh Kell* (Tuscaloosa: University of Alabama Press, 1973), 1-8; U.S. Census, 1850, 1880, McIntosh County.

124. Thomas Marshall Hunter was pastor of the Darien Presbyterian Church from 1894 to 1897, arriving as a young man from Clarkesville, Tennessee with his bride, Sally (Owen) Hunter of Charleston, S.C. Their son, Howard Owen Hunter (1895-1964) was born at the Ridge during Hunter's pastorate in Darien. The younger Hunter was head of the Franklin D. Roosevelt administration's Federal Writers Project of the Works Progress Administration in the late 1930s. His first wife, Mary (Jackson) Hunter (1893-1968) lived most of her life at Cedar Point in McIntosh County. One of their two children, Mary Kate Hunter, married Roy Earl Sullivan, Sr. in 1945. T. M. Hunter is the great-grandfather of the editor of this volume. Sullivan, *Early Days on the Georgia Tidewater*, 515; Sessional records, Darien Presbyterian Church, 1894; White, Index to McIntosh County Cemetery Records, 42; Sullivan-Hunter Family Records.

125. Jesse Alexander Britt, born at Macon, Georgia, May 21, 1864 and died at Darien, Georgia, June 20, 1939, being buried in St. Andrew's Cemetery. J. A. Britt would become J. G. Legare's son-in-law in October 1896 by marrying Claudia Legare (1875-1962). Their children were: Charlotte Legare Britt (1898-1958, later Massey), Emily Wilkinson Britt (1899-1996, later Varnedoe), John Legare Britt (1902-1993) and Jesse Eugene Britt (1904-1974). J. A. Britt was a Darien store owner and was active in the civic and social affairs of Darien and the Darien Presbyterian Church. His business was on the southwest corner of Broad and Screven streets. For a number of years, Britt was ordinary of McIntosh County. He also served as mayor of Darien, 1916-18, and was a city alderman prior to that. He was elected a ruling elder of the Presbyterian Church in 1931. *Darien Gazette*, December 14, 1901, January 20, 1912; Sessional records, Darien Presbyterian Church, 1931; Sanborn Insurance Map of Darien, 1885; White, Index to McIntosh County Cemetery Records, 10; Interviews with Annie Fisher Gill of Darien, January 8, 1997, January 16, 1997.

126. Legare remained in the service of the church as ruling elder and clerk of Session.

127. The Darien and Western Railroad was a new corporation, which had taken over the operation of the defunct Darien Short Line. Track was completed through the eastern section of McIntosh County (roughly paralleling the route of present state highway 99) into Darien in January 1895, thus giving the town its first rail link with the

rest of Georgia. *Darien Gazette,* January 26, 1895.

128. Darien Junction, known earlier as Theo, was in the pine woods section of western McIntosh County, several miles north of Townsend. It was at the juncture of two railroads—the north-south Florida Central & Peninsular (later Seaboard Air Line Railway) and the east-west Darien & Western short line. Darien Junction ca. 1915 became known as Warsaw, which eventually became an active sawmilling community during the 1920s and 1930s. Nothing remains of this community as most activity there had ended by ca. 1940. Sullivan, *Early Days on the Georgia Tidewater,* 579-86.

129. Formed in 1888 from the merger of the Hilton Timber and Lumber Company of Darien and the Dodge interests at St. Simons Island, Georgia, becoming then the largest timber concern on the U.S. Atlantic coast (see note 28). The officers of the Hilton-Dodge Lumber Co. in the 1890s were Joseph Hilton, president, Robert H. Knox, general manager, James Lachlison, superintendent of mills, Robert P. Paul, secretary and treasurer and R. D. Fox. Hilton Papers, No. 387, 1:5, 1:8, 1:11, 2:12 for documents related to the Hilton-Dodge Lumber Company, Georgia Historical Society, Savannah.

130. Legare was to plant rice at Champneys Island for the next eight years. (See the Introduction to this volume).

131. On the second floor of the Darien City Hall building, constructed in 1884. The second floor was added in 1895.

132. Thomas A. Bailey (1854-1919), postmaster of Darien in the 1890s. He was a builder and contractor, store owner and civic leader for many years. T. A. Bailey served a term as sheriff of McIntosh County, 1891-95. He was mayor of Darien 1905-09 and 1910-16 and was chairman of the McIntosh County commissioners in 1919 (he died in office). Bailey also was a city alderman from the late 1890s until his election as mayor of Darien. "He was one of my oldest friends...He was an upright man," Legare wrote in his Journal upon Bailey's death. *Darien Timber Gazette,* July 26, 1890; Howard's Directory, 1892; *Darien Gazette,* October 15, 1898, July 2, 1910; U.S. Census, 1900, McIntosh County; White, Index to McIntosh County Cemetery Records, 8.

133. Albert E. Dimmock, Darien druggist and close friend of J. G. Legare. A. E. Dimmock was a partner of Dr. Peter S. Clark in the Darien firm of P. S. Clark & Co. In 1890 he was elected a ruling elder of the Darien Presbyterian Church. Dimmock moved away from Darien in 1901. He died in 1916 in Valdosta, Georgia. Sessional records, Darien Presbyterian Church, 1890; Howard's Directory, 1892.

134. Actually called towboats locally. They were used for towing drifts of timber and lumber from the booms around Darien to the sawmills at Lower Bluff, May Hall, Union Island and Doboy, as well as to ships loading at Doboy and Sapelo sounds. The Darien & Sapelo Towboat Company had a monopoly on the local business in the 1890s and early 1900s. Sullivan, *Early Days on the Georgia Tidewater,* 476-77, 491-92.

135. Jackson M. Holmes (1829-1901), South Carolina native and Confederate veteran who moved to McIntosh County ca. 1875. He is buried in St. Andrew's Cemetery. U.S. Census, 1880, 1900, McIntosh County; White, Index to McIntosh County Cemetery Records, 23.

136. Charles Joseph McCosker (1856-1898) was survived by his wife Annie Cannon McCosker (1856-1916). Lower Bluff was the largest of the local sawmills, operated by the Hilton-Dodge Lumber Company. *Darien Gazette,* June 25, 1898.

137. The hurricane and tidal wave of October 2, 1898 struck McIntosh County a direct blow, the eye moving ashore in the vicinity of Sapelo Island on a high, spring, tide, causing severe flooding in McIntosh and Glynn counties. *Darien Gazette,* October 8, 1898, October 15, 1898; Sullivan, *Early Days on the Georgia Tidewater,* 511-14.

138. Sarah Ann E. Cobb of McIntosh County (1853-1898) married Lowndes W. Poppell in 1895. She and her husband were serving as caretakers for the Wolf Island Club at the time of the hurricane. Both were lost, as were two children of Sarah Cobb, who may have been born from a previous marriage. Alex Stokely (born 1850) and his wife, Mary (Cobb) Stokely (1858-1898, sister of Sarah Cobb), were either living or visiting on Wolf Island at the time of the storm. The U.S. Census of 1900 for McIntosh County shows that Alexander Stokely was living at the Wolf Island club house and had remarried. Poppell Family Papers; U.S. Census, 1850, 1860, 1880, McIntosh County; *Darien Gazette,* October 8, 1898.

139. Legare kept his journal, along with personal papers and other records at his Champneys Island office.

140. The hurricane of 1898 dealt a devastating blow to Darien's timber industry. Lumber gathered at the river booms was scattered for miles and local sawmills were damaged. In addition, facilities of the timber loading grounds at Front River, Sapelo Sound, were damaged. The various timber firms in Darien engaged every available hand to gather up the widely dispersed timber and lumber scattered in the salt marshes and tidal creeks around Darien and that section of McIntosh County. *Darien Gazette,* October 15, 1898.

141. Lucius Ross Lynn (born 1875) arrived as stated supply for the Darien Presbyterian Church in July 1898 and was installed as permanent pastor on January 18, 1899, Darien being his first pastorate. He resigned in September 1901 to accept a charge at the Rome, Georgia, Presbyterian Church. L. Ross Lynn, with his Darien pastorate, embarked on a long and fruitful career of service to the church, including an extended tenure (1918-1943) as director of the Thornwell Presbyterian Orphanage and pastor of the Thornwell Presbyterian Church in Clinton, South Carolina. In 1961, at the age of 85, Lynn returned to the Darien Presbyterian Church as a guest speaker. Sessional records, Darien Presbyterian Church, 1898, 1901; *Darien News,* January 26, 1961.

142. Henry S. Ravenel, born in 1848 in South Carolina, came to Darien ca. 1875. He was a land surveyor, served as inspector of timber and lumber in the 1880s and 1890s and was a member of the Darien board of pilot commissioners. In 1890, Ravenel was inspector general of timber and lumber for Darien. U.S. Census, 1880, McIntosh County; *Darien Timber Gazette,* May 26, 1882, July 26, 1890; Howard's Directory, 1892.

143. This is probably a reference to George W. Faries (1828-1883) a Savannah native who moved to Darien ca. 1855. He was a commissioner of the McIntosh County Academy from 1866-1875. In the early 1880s, G. W. Faries was inspector general of timber for Darien. William Hunter, with his partner, Robert Manson, was affiliated

with the British firm of Hunter, Benn & Company of Mobile, Alabama, one of the largest of Darien's timber concerns. Hunter, Benn was the company formerly operated as the James Hunter Timber Company in the early 1880s. U.S. Census, 1860, 1880, 1900, McIntosh County; *Darien Timber Gazette,* May 26, 1882; Howard's Directory, 1892. For the fire which destroyed the Presbyterian church, see *Darien Gazette,* July 8, 1899; Sessional records, Darien Presbyterian Church, 1899.

144. Charles Reade Walker (1870-1954), an Ashantilly neighbor and close friend of J. G. Legare. He was a member of the Darien Presbyterian Church and began a long term of service as a ruling elder in 1903. For many years, C. R. Walker was the wireless operator at the Darien Telegraph Company. Sessional records, Darien Presbyterian Church, 1903; Howard's Directory, 1892; White, Index to McIntosh County Cemetery Records, 44.

145. Allen D. Candler, Governor of Georgia, 1898-1902.

146. For insights on the racial unrest in Darien in the summer of 1899, see the Introduction to this volume. Also, W. Fitzhugh Brundage, "The Darien Insurrection of 1899: Black Protest During the Nadir of Race Relations," *Georgia Historical Quarterly* 74 (Summer 1990), 234-53; Sullivan, *Early Days on the Georgia Tidewater,* 517-18; *Darien Gazette,* August 26, 1899.

147. Joseph E. Townsend (1856-1899), son of William Riley and Dorinda Townsend of McIntosh County, Georgia. He is buried in the Townsend family plot at Ebenezer Cemetery, eight miles north of Darien off the coastal highway. Joseph E. Townsend's brother, William R. Townsend, Jr., was, the next day, a leader of the manhunt in northern McIntosh County for those involved in this incident. White, Index to McIntosh County Cemetery Records, 44.

148. Rough rice has not been pounded, i.e. the husks are still attached, as opposed to clean rice, which has had the husks removed. See the Introduction to this volume.

149. Later Ludowici, Long County, Georgia.

150. Charles Odingsells Screven Mallard (1843-1915), McIntosh County rice planter and civic leader. He cultivated rice at Oasis on Cathead Creek west of Darien in the late 1800s and early 1900s. C. O. S. Mallard was the son of Thomas Samuel Mallard (1816-1882) of Liberty County, Georgia, who acquired Oasis in McIntosh County from the Gignilliat family in 1852. C. O. S. Mallard's wife was Minnie (Gignilliat) Mallard (1845-1913). He served as a ruling elder of the Darien Presbyterian Church from 1900 until his death in 1915. Sullivan, *Early Days on the Georgia Tidewater,* 219-23; U.S. Census, 1850, Liberty County; Howard's Directory, 1892; White, Index to McIntosh County Cemetery Records, 38; Sessional records, Darien Presbyterian Church, 1900.

151. Train depot at Columbus Square in Darien, marking the terminus of the Darien & Western line.

152. The rice tracts on Cathead Creek were usually the most susceptible to severe flooding when freshets occurred in the Altamaha River.

153. Catherine Cromley (ca.1830-1900) was the wife of James Cromley (ca.1809-1890), both being born in Ireland. The family arrived in Darien from an earlier residence in Savannah ca. 1855. James Cromley, after the Civil War, became the keeper of the Sapelo Island lighthouse and Catherine Cromley was assistant keeper. The

Cromleys and their sons lived on the lighthouse tract for over sixty years, until lighthouse operations were ended on Sapelo Island by the federal government in 1933. Their sons also served as keepers of Sapelo light and the Wolf Island range light. Sullivan, *Early Days on the Georgia Tidewater*, 417-18, 424; U.S. Census, 1860, 1870, McIntosh County; White, Index to McIntosh County Cemetery Records, 13.

154. Tabby, comprised of equal parts of sand, water, oyster shell and quicklime, was a popular building material used primarily during the antebellum period for the construction of sugar mills, barns and other plantation structures. Extensive tabby ruins are still in evidence at Darien, the Thicket and Sapelo Island. The most cogent discussion of tabby and its applications is contained in Floyd, "Certain Tabby Ruins on the Georgia Coast," in *Georgia's Disputed Ruins*, 3-189; also, Sullivan, *Early Days on the Georgia Tidewater*, 101-07, 766-68, 824-27.

155. The Presbyterian Church which burned July 6, 1899 was built on Bayard Square in 1876. The new church on the same site completed in the summer of 1900 is the present edifice.

156. Legare's stepson. He died July 29, 1900 and is buried in St. Andrew's Cemetery. See the Introduction to this volume.

157. These were the workers who loaded timber and lumber aboard the vessels at two Front River docks at the west end of Sapelo Sound. They were operated by the Hunter, Benn and Company and J. K. Clarke timber firms of Darien. Hilton-Dodge had a similar facility across the sound on the Julianton River near the lower end of Harris Neck. Many of these African-American longshoremen lived on the north end of nearby Creighton Island, convenient to the loading grounds. U.S. Census, 1900, McIntosh County; Sullivan, *Early Days on the Georgia Tidewater*, 488-98; U.S. Coast and Geodetic Survey, "Description and Hydrological Report for Sapelo Sound, Georgia," Washington, D.C., 1902.

158. The year 1900 was the most active in the history of the Darien timber and lumber industry as a record amount of timber and lumber, 112.5 million linear board feet, was shipped on foreign and domestic vessels. After 1900, timber shipments from McIntosh County steadily declined until, by the start of World War I, the industry had all but ended. *Darien Gazette*, January 19, 1901; Morrison, "Raftsmen of the Altamaha" (Unpub. masters thesis, University of Georgia, 1970), 81-97; Sullivan, *Early Days on the Georgia Tidewater*, 535-72.

159. Because they were strike-breakers.

160. See Sullivan and Haynes, *History of the First Presbyterian Church of Darien, 1736-1986.* 29-31.

161. S. A. Way and Alexander Campbell Wylly (1833-1911). S. A. Way was clerk of McIntosh County Superior Court throughout the 1890s until being defeated by A. C. Wylly for the office in 1900. Wylly was inspector of timber and lumber for Darien during the 1880s and 1890s, and was surveyor of Darien and McIntosh County for many years. *Darien Timber Gazette*, July 26, 1890; *Darien Gazette*, December 14, 1901; Howard's Directory, 1892; White, Index to McIntosh County Cemetery Records, 46.

162. Peter Stratton Clark (1857-1919), Darien physician and druggist for many years. He lived at the Ridge and conducted his medical practice from the same building on Broad Street in Darien in which his drug store was located. P. S. Clark was in

Darien as early as 1882 when he is listed in the *Darien Timber Gazette* as port physician. For many years, the P. S. Clark & Co. drug store, "Dealers in drugs, patent medicines, oils, paints, school books, stationery, toilet articles, golf balls and clubs, base ball goods and garden seeds," was one of the most frequented local stores with its variety of services. Dr. P. S. Clark was one of the most beloved Darien citizens of the period. He was married to Maria (Lachlison) Clark (1859-1941). Both are buried in St. Andrew's Cemetery. *Darien Timber Gazette,* May 26, 1882; *Darien Gazette,* April 13, 1907, June 12, 1909; Howard's Directory, 1892; White, Index to McIntosh County Cemetery Records, 12.

163. For many years Legare kept the minutes of the Darien Board of Pilot Commissioners, which regulated the activities of the branch (bar) pilots who guided shipping into and out of the local harbors of Doboy and Sapelo sounds. The Board also established rates and rules of pilotage and arbitrated disputes between the pilots—it being an highly competitive business. Minutes of the Board of Pilot Commissioners, 1880-1930, City of Darien Archives; Sullivan, *Early Days on the Georgia Tidewater,* 457-61.

164. Fortunato Nanias, Darien merchant. His store was on Broad Street. Howard's Directory, 1892.

165. I. Joselove, Darien merchant. He owned a dry goods store on the north side of Broad Street just east of its intersection with Screven. He later changed his name to Josel and his clothing store became known as "A. Josel, Darien." *Darien Gazette,* April 13, 1907, January 27, 1912.

166. Meyer Bluestein (1870-1941), Darien merchant and grocer from the 1890s until well into the 1900s. He arrived in the United States from Russia in 1885 as a boy of fourteen. He is thought to have come to Darien ca. 1895. A brother, Joseph Bluestein (note 189) also came to Darien. In Darien, Meyer Bluestein ran a store on the northwest corner of Walton and Broad streets called the Shipwreck. Later, he owned and operated one of Darien's most successful businesses in the brick store at the corner of Walton and Broad on the site of his earlier store. The *Darien Gazette* of January 15, 1910 reveals that Meyer Bluestein was also engaged in the commercial harvest of oysters as a news item notes the loss by fire of his oyster boat at Harris Neck. Meyer Bluestein was married to Minnie Asman (1877-1956), also of Russian birth. Her brother, B. Asman, was a tailor and opened a clothing business in Darien in 1900. Several of the children of Meyer and Minnie Bluestein were also active citizens of Darien. Annie Bluestein (1893-1966) married Emanuel Hackel (note 216) who ran a business on Walton Street adjacent to Meyer Bluestein's grocery store. Jack Bluestein (1911-1994) married Florence Johnson and ran his father's grocery business on Walton Street for many years. In the early 1930s, Jack Bluestein ran this business with his brother, Walter Bluestein (1906-1955), before taking over on his own. Another son, Sam Bluestein (1912-1993) married Hazle Tankersley and, for many years, operated a service station directly across the street from Jack Bluestein's grocery store, on the northeast corner of Walton and Broad. *Darien Gazette,* December 18, 1897, December 17, 1898, April 1, 1899, August 18, 1900, May 26, 1933; Bluestein Family Records; Interview with David Bluestein, July 6, 1997.

167. Florence Calder Lachlison (1859-1947), wife of James Addie Lachlison (1861-

1891), was a teacher in the McIntosh County schools for many years in the late 1800s and early 1900s. Florence Lachlison gathered and recorded much of the history of McIntosh County during her life and was greatly involved in the civic, cultural and social activities of Darien. White, Index to McIntosh County Cemetery Records, 26.

168. F. H. McFarland was president of the Darien & Western Railroad. From 1903-05 he was a director of the Darien Bank. McFarland served two terms as chairman of the McIntosh County Board of Commissioners, 1905-12 and 1913-19.

169. See note 84.

170. Lectured Crawford, a black legislator, was a McIntosh County representative in the Georgia legislature during the postbellum period. Crawford was one of a number of African-Americans who were prominent in the powerful black political machine begun in McIntosh County by Tunis G. Campbell (note 33), which predominated locally from the early 1870s until 1907. Crawford served several terms as a representative in the state legislature: 1886-87, 1890-91 and 1900-01. Other African-Americans from McIntosh County who served in the legislature during this period were Amos Rogers, 1878-79, Hercules Wilson, 1882-85, and W. H. Rogers, 1902-07. On the county level, a number of blacks had positions of authority, including Lewis Jackson, who served as county ordinary throughout the 1870s, James Bennett, who was county sheriff during the same period, and Alonzo B. Guyton, deputy sheriff and constable in Darien for many years (note 27). In the 1890s, other African-Americans were prominent locally, including Charles R. Jackson, postmaster of Darien, John C. Lawton, deputy collector of customs for the port of Darien, Isaac Polite, Darien grocer, Baptist ministers E. M. Brawley, Paul R. Mifflin and J. P. Davis, Presbyterian minister J. D. Taylor, G. W. Butler of the African Methodist Episcopal Church, F. M. Mann of St. Cyprian's Episcopal Church, S. W. McIver, chairman of the Republican Party and James L. Grant, educator and editor of the *Darien Spectator.* Duncan, *Freedom's Shore,* 98, 100, 114-16; *Darien Timber Gazette,* October 3, 1874; Howard's Directory, 1892.

171. Hudson was a small African-American community eight miles northeast of Darien on the Cow Horn Road. Its outlet to Doboy Sound was by way of Hudson Creek. The railroad had a scheduled stop at Hudson on its daily runs.

172. Townsend was established ca. 1885 in the pine woods of western McIntosh County by Joseph E. Townsend, while he was obtaining railroad rights-of-way for the Florida Coast & Peninsular Railroad which was eventually built through coastal Georgia in the early 1890s. This railroad later became the Seaboard Air Line Railway. Townsend was a mail and passenger stop for the Seaboard for many years during the era of timbering and naval stores production in that section of McIntosh County (see also note 253). Sullivan, *Early Days on the Georgia Tidewater,* 575-78, et al.

173. For the rice mill boiler.

174. Robert Manson (1865-1926) and his brother John Manson (1869-1928), both natives of Scotland, came to Darien in the late 1880s to manage, with William Hunter (note 143), the Hunter, Benn and Company timber firm of Mobile, Alabama. Robert Manson was local manager of the concern in the 1890s. The Scots brothers are also said to have introduced the game of golf to Darien and were instrumental in starting the Darien Golf Club in the 1890s with the links being laid out approximately in the area of the present high school site. The Mansons and their children were all avid

golfers. The first club house of the Darien Golf Club is still standing. This structure is now a private residence at Valona. Lewis, *They Called Their Town Darien* (Darien, Ga.: The Darien News, 1975), 59; Howard's Directory, 1892; White, Index to McIntosh County Cemetery Records, 28.

175. Isidore Collat of Savannah operated a dry goods business in Darien for many years. Born in Germany in 1858, he came to Darien from Savannah as a young man in the late 1870s. Isidore Collat later moved back to Savannah then, as noted by Legare in his Journal, returned to Darien in June 1903 to work in his store. His store in the 1880s, which he ran with his brothers, A. H. Collat and Louis Collat, sold clothing for men, women and children, boots and shoes, jewelry, furniture and "Buggies & Wagons on the installment plan," according to one newspaper advertisement. The Collats also had a store in Brunswick in the 1890s. *Darien Timber Gazette*, April 14, 1883, September 29, 1888; Howard's Directory, 1892.

176. William Henry Strain (1875-1924), Darien merchant, businessman and rice planter. He was the son of Adam and Emma Strain of Darien (note 8). In the late 1890s and early 1900s, W. H. Strain was the senior member of the Darien firm of Adam Strain's Sons. With his younger brother Robert (note 194), he leased Butler's Island and planted rice there, later purchasing the island. *Darien Gazette*, July 8, 1899; White, Index to McIntosh County Cemetery Records, 42.

177. Daniel Cromley (born 1861), assistant keeper of the Sapelo Island lighthouse in the early 1900s. He was the son of James and Catherine Cromley (note 153). Daniel Cromley, like his father, was a cobbler and bootmaker. His brother, James Cromley (1865-1940), was the keeper of the lighthouse in 1903. U.S. Census, 1880, 1900, McIntosh County; White, Index to McIntosh County Cemetery Records, 13.

178. At this time, W. H. Strain was planting rice on Butler's and Generals islands (leased from Frances Butler Leigh and Sarah Butler Wister). In 1906, Strain exercised an option to purchase Butler's Island, then acquired Generals Island in 1907. The Butler family had owned Butler's Island for well over a century.

179. James Dean (1829-1903), McIntosh County bar pilot and waterman, for many years lived on the Ridge. He operated sail-driven pilot boats before, during and after the Civil War, guiding timber ships in and out of Doboy Sound. His best-known boat was the *Young America*, one of the fastest sailers on the Georgia coast. Sullivan, *Early Days on the Georgia Tidewater*, numerous references; White, Index to McIntosh County Cemetery Records, 15.

180. Cedar Point, eleven miles northeast of Darien off the Cow Horn Road, fronting the salt marsh overlooking Creighton and Sapelo islands. Cedar Point was the antebellum Sea Island cotton plantation of Henry S. Atwood and the postbellum residence of one of his sons, William H. Atwood (note 31). The Atwood home on Cedar Creek, built in 1890 by W. H. Atwood, still stands. An oyster cannery was later located at Cedar Point. Sullivan, *Early Days on the Georgia Tidewater*, 493, 534, 687, et al.

181. Edward R. Poppell (1840-1903), western McIntosh County businessman and for many years engaged with other members of his family in the naval stores industry. White, Index to McIntosh County Cemetery Records, 34.

182. Apparently, King's recruiting mission was not successful as Legare makes no further mention of it.

183. It should be remembered that no roads connected the islands and plantations of the Altamaha River delta with Darien and the mainland even in the years just after the turn of the century. All travel was by boat. Legare, when he was not staying on Champneys Island, traveled by boat from the Darien waterfront, or from Snow Creek at his Ashantilly residence, to Champneys via Generals Cut, which ran through the middle of Generals Island connecting the Darien and Butler rivers. Three Mile Cut was further east, between Rhett's and Rockdedundy islands. It was largely used by towboats transporting drifts of pine timber from the Darien booms to the St. Simons Mills. Sullivan, *Early Days on the Georgia Tidewater*, contains details on the various cuts and canals around Darien.

184. An acute virile disease resembling influenza.

185. Georgia Tech in Atlanta.

186. Formerly Jonesville (note 43). In 1904, Jones Station was on the Seaboard Air Line Railway which passed through western McIntosh County. Sullivan, *Early Days on the Georgia Tidewater*, 586-88, et al.

187. Ann Harvey Shine (1817-1905), wife of Daniel Yates Shine (1805-1874). D. Y. Shine was a McIntosh County businessman and planter prior to the Civil War. White, Index to McIntosh County Cemetery Records, 39.

188. Robert Austin Young, Sr. (1866-1967), receiver of tax returns for McIntosh County and Darien civic leader for many years. He served as receiver of tax returns almost continuously from 1893 through 1928. R. Austin Young, Sr. also served on the Darien aldermanic council for several years and was elected mayor in November 1929, serving an eight-year term through 1937 and a second term from 1940-42. John G. Legare's Journal entry of July 3, 1901 relates to the unfortunate death of Young's first wife, Rosalie (Gardner) Young (1876-1901). He was later married to Ruth (Middleton) Young (1886-1941). His son, R. A. Young, Jr. (1900-1988) served McIntosh County as tax commissioner from 1929 to 1965. Darien Aldermanic Minutes, 1921, 1929-30, City of Darien Archives; *Darien Gazette*, April 13, 1907, June 24, 1911; Interviews with Lloyd (Young) Flanders, March 10, 1997, March 18, 1997.

189. A clipping from the *Savannah Morning News* attached to this page of Legare's journal relates the sad death of fifteen-year old Hyman Bluestein, son of Joseph Bluestein of Darien and Rosa Bluestein (died 1900). Bluestein and a black youth, Preston Middleton, were drowned in the Altamaha River after their small bateau capsized near Darien. Funeral services for the Bluestein lad were held at the residence of Mrs. Meyer Bluestein in Savannah and he was buried in Laurel Grove Cemetery. Joseph Bluestein (b. 1870), native of Russia, was the brother of Darien merchant Meyer Bluestein (note 166). *Darien Gazette*, April 8, 1905.

190. The Shell Bluff Canning Company was located on the banks of Shellbluff Creek at Valona, ten miles northeast of Darien. Shell Bluff was a small community established by members of the Atwood family on lands of the former Manchester plantation. In 1898, Shell Bluff's name was changed to Valona for post office purposes. The name, ostensibly given by George Atwood, was for the port of Valona, Albania, from which place a ship was in local waters to load lumber. The Shell Bluff Canning Company, managed by the partnership of James L. Atwood of Valona and Robert A. Strain of Darien, was engaged in the cultivation and harvest of oysters in McIntosh

County tidal waters. Oysters brought in were canned and distributed from the factory at Valona. *Darien Gazette,* July 22, 1905; Sullivan, *Early Days on the Georgia Tidewater,* 686-87, 691, et al.

191. A white tasteless compound used in medicine as a purgative and fungicide.

192. This was to facilitate shipments of yellow pine timber from the interior of Georgia to Hilton & Dodge's Lower Bluff sawmill at Darien. By 1905, the timber industry was in decline at Darien. The railroad came too late to Darien to save the industry. William M. Kelso, "Excavations of the Fort King George Historical Site," Georgia Historical Commission, 1968; U.S. Army Corps of Engineers survey map, Darien Quadrangle, 1918.

193. This was never done. The south end of Black Island is directly across the marsh from Lower Bluff and fronts on the broad stretch of the Darien River (north branch of the Altamaha) known locally as Long Reach.

194. Robert Adam Strain (1879-1928), son of Adam and Emma Strain of Darien (note 8), and younger brother of William H. Strain (note 176). The Strain brothers were enterprising Darien businessmen, operating the general store begun by their father. U.S. Census, 1880, McIntosh County.

195. Charles A. Dolbow bought and leased tidal salt marshes in McIntosh County in the early 1900s for the purpose of oyster harvesting. Sullivan, *Early Days on the Georgia Tidewater,* 716.

196. James Alfred Atwood (1840-1906) of Shell Bluff (Valona), son of Henry S. and Ann (McIntosh) Atwood, and one of four brothers, the others being William Henry (note 31), John M. (note 213) and George E. (note 221). U.S. Census, 1880, 1900, McIntosh County.

197. William A. Wilcox (1839-1907), Darien merchant and businessman. He was co-owner of a store with William W. Churchill (note 11). In addition, Wilcox was proprietor of the Club Saloon in Darien in the 1880s and 1890s. Wilcox married Churchill's daughter, Edytha Churchill (1846-1918). The Churchills and Wilcoxs are all interred in a marble vault, one of the most impressive structures in St. Andrew's Cemetery. U.S. Census, 1880, McIntosh County; Howard's Directory, 1892; White, Index to McIntosh County Cemetery Records, 12.

198. Arthur Bailey (1836-1910), born in Ireland, resident of the Ridge (Ridgeville) with his wife, Mary Ann Bailey (1839-1916), from the 1860s. Bailey, a merchant and saloon owner, came to McIntosh County in the late 1850s. U.S. Census, 1860, 1900, McIntosh County; Howard's Directory, 1892; White, Index to McIntosh County Cemetery Records, 7.

199. Daniel Legare Whitesides (1875-1961), born in Mt. Pleasant, South Carolina, near Charleston. He served in the Spanish-American War and came to Darien ca. 1900 to work for J. G. Legare. He was married for fifty-three years to Legare's daughter, Emma Legare Whitesides (1885-1973). Their children were Joseph Legare Whitesides (born 1908) and Claudia Whitesides (1911-1983). About 1911, the Whitesides family acquired the Samuel R. Dean house in Darien across from the City Hall building, and lived there for the remainder of their lives. U.S. Census, 1900, McIntosh County; Interview with Mary Sue Murray, January 12, 1997.

200. One of the lengthiest entries recorded by Legare in his role as clerk of ses-

sion of the Darien Presbyterian Church relates to the circumstances involving Smith's resignation from his pulpit. The pastoral severance between Smith and the Presbyterian session originated over a disagreement relating to vacation time for the former. Sessional Records, Darien Presbyterian Church, October 6, 1907, February 2, 1908.

201. William S. Milne served as pastor of the Darien Presbyterian Church from 1908-1911.

202. Augusta Powell Pease (1827-1909) was the widow of Darien businessman Theodore P. Pease (note 15). Both are buried in St. Andrew's Cemetery. White, Index to McIntosh County Cemetery Records, 34.

203. James Hines (1860-1909) of Hinesville, Georgia, county seat of Liberty County, was the son of Charlton and Sarah Jane (Way) Hines. The elder Hines was a Liberty County state senator from 1828 to 1843 and was the founder of Hinesville. The Georgia Coast & Piedmont Railroad was the successor of the Darien & Western. This shortline road ran from Collins, Georgia, through Tattnall, Liberty and McIntosh counties to its terminus in Darien. Robert Manson Myers, *The Children of Pride* (New Haven: Yale University Press, 1972), 1550.

204. Fred Stewart Leigh Grundy (1886-1924) and Louise Hopkins Grundy (1881-1980), both buried in St. Andrews Cemetery. White, Index to McIntosh County Cemetery Records, 22.

205. Francis E. Durant (1861-1910) of Meridian, McIntosh County, was treasurer of McIntosh County from the late 1880s to the early 1900s. Meridian was on the Cow Horn Road and a stop on the Georgia Coast & Piedmont Railroad route seven miles northeast of Darien. The home of the Durant family, a post office was opened there in 1896. *Darien Timber Gazette*, September 29, 1888; *Darien Gazette*, August 17, 1895, December 14, 1901; White, Index to McIntosh County Cemetery Records, 18; for Meridian, Sullivan, *Early Days on the Georgia Tidewater*, numerous references.

206. William Screven Mallard (1845-1910), local rice planter.

207. John Hagan (1847-1910) of the Ridge was a local bar pilot in Doboy and Sapelo sounds during Darien's timber boom. U.S. Census, 1900, McIntosh County; White, Index to McIntosh County Cemetery Records, 22.

208. Samuel Randolph Dean (1865-1910) managed a meat market in Darien in the 1890s and early 1900s and was a city alderman for several years. He was married to Frances (Young) Dean (1872-1959). Both are buried in St. Andrew's Cemetery. Howard's Directory, 1892; Darien Aldermanic Minutes, 1906-07, City of Darien Archives; White, Index to McIntosh County Cemetery Records, 16.

209. Sidney Walter Williams (1854-1911) of South Newport was buried in Baker Cemetery off Harris Neck Road near the South Newport River. White, Index to McIntosh County Cemetery Records, 46.

210. John Philip Mew (1879-1945) was an insurance agent with an office in Darien. A native of Hampton, South Carolina, he was married to Louise Sinclair of Darien, daughter of William W. and Lillian (Clark) Sinclair. Louise S. Mew (1888-1974) was a traveling clothing merchandiser. See also JGL Journal entry of December 10, 1913. White, Index to McIntosh County Cemetery Records, 29.

211. William Clifton (1854-1911) of the Ridge was state senator from McIntosh

County in the 1890s and early 1900s. *Darien Gazette,* March 31, 1900; White, Index to McIntosh County Cemetery Records, 12.

212. Carl August Schmidt (1837-1914) was one of Darien's most enterprising timber brokers. He began marketing timber and lumber in Darien ca. 1874 and was a local mainstay in the trade through the first decade of the 1900s. Schmidt was German vice-consul for Darien in the 1880s and 1890s. The observation here by Legare on the failure of Schmidt's firm is further evidence of the decline of Darien's timber fortunes. *Darien Timber Gazette,* May 26, 1882; U.S. Census, 1880, McIntosh County; White, Index to McIntosh County Cemetery Records, 38.

213. John M. Atwood (1849-1933) was the son of Henry S. and Ann McIntosh Atwood and the twin brother of George E. Atwood (1849-1914, note 221). John M. Atwood operated a store and the post office at Crescent near the railroad depot.

214. When John G. Legare sold his last crop on Champneys Island in late 1904, the local rice industry was already in severe decline. Very little commercial rice was grown in the Darien region after 1905. See the Introduction to this volume. In 1929, a soil map of McIntosh County was compiled by the U.S. Department of Agriculture. It delineated the Coastal Agricultural Experiment Substation, managed by the state of Georgia, on Butler's and Champneys islands, on either side of the old track bed of the G.C. & P. railroad. Farming operations here were being revived by T. L. Huston of New York, owner of the islands at that time, and geared to the cultivation of truck crops on lands where rice had been grown a generation before. Although not mentioned by Legare in his Journal, he undoubtedly followed these activities with considerable interest. G. L. Fuller, B. H. Hendrickson and J. W. Moon, *Soil Survey of McIntosh County, Georgia,* Series 1929 (Washington, D.C.: U.S. Department of Agriculture, 1930).

215. Legare is here commenting on the start of work by the Georgia Coast & Piedmont Railroad to extend its track southward from Darien to Brunswick by constructing trestle work and steel bridges over the marsh, rice fields and tidal streams of the Altamaha delta. The work took about two years, being completed in 1914. Sullivan, *Early Days on the Georgia Tidewater,* contains documentation and photographs of this work, which involved considerable labor and engineering skill.

216. Emanuel Hackel (1886-1975) came to Darien as a young man ca. 1912 and was for many years through the 1960s a local merchant and city resident. In the 1920s, he and his wife, Annie (Bluestein) Hackel (note 166), resided on Sapelo Island where they were employed by the island owner, Howard E. Coffin. He ran the commissary at Barn Creek and Mrs. Hackel was the island postmaster. When the Hackels returned to Darien, Emanuel Hackel ran the package store on Walton Street adjacent to the Bluestein grocery store for many years. Bluestein Family Records.

217. This building was placed beside the G.C. & P. track overlooking the Darien River, precisely where the present Darien News building now stands. The original structure burned in 1971. The new steel swing-span bridge, when completed in March 1914, crossed the river in the same place that the present Highway 17 bridge was built in 1944.

218. Reverend F. M. Mann was the rector of St. Cyprian's Episcopal Church in Darien. His wife, Mary Alexander Mann, established the Mann School for black children at St. Cyprian's Church in the late 1880s with the assistance of Frances Butler

Leigh. Mary Mann was the daughter of Aleck and Daphne Alexander who were Butler's Island slaves. F. M. Mann received financial assistance for defendant Henry Delegal from Sarah Butler Wister during the racial troubles in Darien in August-September 1899. Bell, *Major Butler's Legacy*, 559-60.

219. The passenger depot for the G.C. & P. in Brunswick was built at the corner of Gloucester Street and Cochrane Avenue. Collins, in Tattnall County, was the other end of the G.C. & P. line from Brunswick, a track distance of ninety-nine miles. From Darien northward, the G.C. & P. followed a route which roughly paralleled the Cow Horn Road through Ridgeville, Carneghan, Meridian and Hudson. There were spur stops to serve Valona and Cedar Point, the latter at Oak Hill. From Oak Hill the train stopped at Crescent, about two miles beyond. Crescent was the site of the machine shops, engine repair works, water tank and other facilities for the G.C. & P. From Crescent, the route went to Eulonia and Darien Junction (Warsaw after ca. 1915), both in McIntosh County, Tibet and Ludowici in Liberty County, and Glennville, Reidsville and Collins in Tattnall. An analysis of railroading activities in Darien and McIntosh County is contained in Sullivan, *Early Days on the Georgia Tidewater*, see index.

220. Hugh Boyd Manson (1891-1958) was a son of Darien timber merchant Robert Manson (note 174). Sarah (Sadie) Clark Manson (1892-1983) was the daughter of Dr. P. S. and Maria (Lachlison) Clark (note 162). After a residence in Florida, the Mansons resided on the bluff at Cathead Creek in Darien for many years. Interview with Annie Fisher Gill, March 20, 1997.

221. George Elliott Atwood (1849-1914), son of Henry S. and Ann (McIntosh) Atwood of McIntosh County, lived at Valona, formerly Shell Bluff, where he was a merchant and ship chandler for many years. He served as a McIntosh County commissioner in the 1890s and early 1900s and also represented the county in the state house of representatives. He was buried in the Atwood family cemetery at Valona. George E. Atwood was the twin brother of John M. Atwood (note 213). As the reader has by now ascertained, many of the deaths recorded by John G. Legare in his Journal are attributed to Bright's disease, which generally related to various forms of kidney failure. Legare was greatly interested in medical matters. He kept a number of medical books on hand and appears to have frequently consulted them. *Darien Timber Gazette*, March 21, 1891; *Darien Gazette*, November 30, 1901; White, Index to McIntosh County Cemetery Records, 7; Sullivan, *Early Days on the Georgia Tidewater*, numerous references.

222. When Emily Britt Varnedoe died December 9, 1996 she had been a member of the Darien Presbyterian Church continuously for eighty-two years. For many years, she was Ordinary of McIntosh County, just as her father, Jesse A. Britt, had once been.

223. Matthew James Dean (1865-1915), resident of the Ridge, was a McIntosh County bar pilot who, like his father, James Dean (note 179), worked the timber shipping entering and leaving port in Doboy and Sapelo sounds. White, Index to McIntosh County Cemetery Records, 16.

224. J. G. Forbes, Darien merchant and grocer in the early 1900s. His establishment at the corner of Broad and Screven streets specialized in "choicest groceries & canned goods, also hardware, tinware & crockery, hay, grain & country produce." He

was also owner of the Altamaha Woodworking Company, which operated from 1909 to 1919 near Lower Bluff to manufacture wooden handles for tools and farm implements. Forbes was active in local politics and served as chairman of the McIntosh County Commission, 1923-26. He was president of the Darien Bank in 1921-22. *Darien Gazette,* June 12, 1909; Sullivan, *Early Days on the Georgia Tidewater,* 563-64.

225. William A. Brannan (born ca. 1871) operated a store and small boarding house on Walton Street in Darien across from the court house. He was also involved in the naval stores business.

226. Porter Middleton, Darien druggist. He built the brick building in 1916 on Broad Street across from the G.C. & P. depot. Soon after opening his business, Middleton removed to Hazlehurst, Georgia, and James L. (Jim) Stebbins bought the building.

227. See notes 162 and 175.

228. George W. Poppell lived near Cox in southwestern McIntosh County and had a small farm in that section. F. H. McFarland (note 168) was chairman of the county board of commissioners at this time.

229. The Stage Road had been so named in the colonial period soon after the first route had been established between Savannah and Darien. The Stage Road, in the 1920s, became the Atlantic Coastal Highway (US 17).

230. Julian Austin Space (1875-1936) was cashier of the Darien Bank from 1899 to 1913 and again from 1915 to 1921. He was dismissed from his position at the bank in 1921 after allegations of his having made unauthorized transactions, and was replaced in the position of cashier by D. E. Lane (note 271). Darien Bank: Minutes of the Board of Directors, 1921. Sullivan, *The Darien Bank,* 44, 73.

231. A. P. Lee died October 15, 1916 and was buried at White Springs, Florida.

232. Edward Gendron Cain, Jr. (1879-1934), native of Charleston and son of Edward G. Cain, Sr., and Emily Ravenel Cain. The senior Cain (1845-1925) was involved in the local timber trade and was an inspector of timber and lumber for Darien in the mid-1890s. E. G. Cain, Jr. was mayor of Darien in 1923-24 and was chairman of the McIntosh County Commission, 1920-21 and 1927-1934. Darien Aldermanic Minutes, 1894-95, 1923-25, City of Darien Archives; White, Index to McIntosh County Cemetery Records, 11; *Darien Gazette,* August 17, 1895; U.S. Census, 1900, McIntosh County.

233. The position of city clerk of Darien was not an elected office but was, at times, very political as seen by the events of the 1916 local election.

234. Raymond Winfred Clancy (1878-1958), married to Mary (Manson) Clancy (1884-1977), both buried in St. Andrew's Cemetery. Raymond Clancy was Darien postmaster in the 1920s and 1930s when the post office was on the west end of Broad Street.

235. David Sutherland Walker (1887-1927), married to Margaret (Manson) Walker (1887-1929), both buried in St. Andrew's Cemetery.

236. James Lachlison Stebbins (1878-1945), married to Victoria (Gnann) Stebbins (1880-1943), both buried in St. Andrew's Cemetery. Jim L. Stebbins was a public servant for many years and was one of the most respected and well-liked citizens of Darien. He was city alderman for a number of years and served as mayor of Darien, 1925-29.

Stebbins ran a grocery business in his brick store on Broad Street in Darien, a business later run for many years by his son James Robert Stebbins (1907-1978).

237. A beautiful home on the marsh overlooking the Crescent (South Sapelo) River, the lifelong residence of Capt. Thomas Spalding Hopkins (1891-1982) and his wife, Margaret (Webb) Hopkins (1899-1988), both buried in St. Andrew's Cemetery.

238. The Darien Shipbuilding Company was a cooperative venture by several businessmen to inject new life into a sagging local economy. The yard was located on the Darien River on property later occupied by the Ploeger Packing Company. Robert Manson was president of the shipbuilding firm. Robert John Downey (1878-1949) was secretary, a position later held by Charles M. Tyson (note 261). R. J. Downey was chairman of the McIntosh County Commission in 1912-13 and was married to Aline (Paul) Downey (1886-1946).

239. The steamboat *Hessie* was continuing regular service from Brunswick to Darien but its business had been sharply curtailed after the railroad was opened between the two cities in 1914. Brunswick steamboat agent J. B. Wright finally suspended the *Hessie's* runs in 1918. For a brief period of time in 1920 the *Hessie* made runs to Darien once a week. Sullivan, *Early Days on the Georgia Tidewater*, 436-38.

240. The wife of Houstoun Legare (1889-1965) was actually named Louella (Dollie) Wells (1891-1974). Their children were James Houstoun Johnston Legare, Jr. (born 1919), and John Edward Legare (1921-1995).

241. During these years travelers on the G.C. & P. between Darien and Brunswick had the option of utilizing the "automobile transfer" by which their personal automobiles were placed on flat cars on either side of the Altamaha delta and transported across the former rice fields of Generals, Butler's and Champneys islands and Hofwyl-Broadfield on the Glynn County side. The automobile transfer ran from March 1914 to November 1919, when the G.C. & P. went into receivership. Photos of this activity are contained in Sullivan, *Early Days on the Georgia Tidewater*.

242. Mary B. Roughan, native of Limerick, Ireland, died August 27, 1920, aged 74 years, and is buried in St. Andrew's Cemetery. White, Index to McIntosh County Cemetery Records, 36.

243. The Darien Telephone Company was established in 1911. The manager at the time of this incident was E. T. Wilkins (not to be confused with E. G. Wilkins). The Darien Telephone Company was subsequently acquired by Joseph C. Jackson in 1921. Jackson died in 1924, and in 1942, his son and daughter in law, Richard V. Jackson (1909-1985) and Bessie Ridley Jackson (1911-1989), assumed management of the telephone company. In 1960, the business moved to its present location on US 17 in Darien where it continues to be managed by descendants of Joseph Jackson.

244. Dr. W. J. Long lived at Townsend for many years and provided medical service to much of the north section of McIntosh County. He came from Liberty County, the section that became Long County in 1920 and which was once part of McIntosh until 1872.

245. William B. Hagan (1841-1919), married to Anna (Durant) Hagan (1863-1902), both buried in St. Andrew's Cemetery. He was a bar pilot who worked the timber shipping in Doboy and Sapelo sounds from after the Civil War to just after 1900, being the captain of the *Unique* for many years. City of Darien, Records of the Board

of Pilot Commissioners.

246. Samuel J. Hagan (1888-1925), son of William B. and Anna (Durant) Hagan. He was married to Marie L. (Sutton) Hagan (1889-1915).

247. Union Island was just east of Blue and Hall Landing at the Ridge, fronting on North (Ridge) River en route to nearby Hird Island. Union Island for many years was the scene of extensive sawmilling activity and timber and lumber loading. It was the site of two mills owned by the Lachlison family. Union Island was later part of the Hilton Timber and Lumber Company. It was known locally as Pumpkin Hammock and was delineated as such on local maps and navigational charts until the early 1900s. During the postbellum period, there were plank walks through the marsh providing Union Island with access to both Blue and Hall and Hird Island, which was just east of Union. Hird Island had been the scene of extensive lumber sawmilling since pre-Civil War days (see note 29). Both Union and Hird islands suffered heavily in the October 2, 1898 hurricane. At that time, the widow of Charles Joseph McCosker (note 136), Annie Cannon McCosker, lived on Hird Island with her seven children. The tidal wave cost them everything they owned. They moved to the mainland and occupied the former L. E. B. DeLorme home just south of the Ridge. Sullivan, *Early Days on the Georgia Tidewater*, numerous references to Union and Hird islands; Lower Altamaha Historical Society newsletter, March 1997, from information provided by Annie Fisher Gill.

248. Anne Lee Haynes (1896-1997) and Frances Haynes (1898-1989) were young sisters who lived at Ashantilly in the old residence built ca. 1820 as their mainland home by Thomas and Sarah Spalding of Sapelo Island. The sisters were natives of Atlanta. Frances Haynes was librarian at Florida State University for more than thirty years. Their younger brother, William G. Haynes (1908-), continued to reside in the Ashantilly house, restored after a 1930s fire, at the time of this writing in 1997.

249. Charles L. Bass (1853-1919) of Sapelo Island, cousin of Thomas, Thomas Bourke and Sarah E. Spalding McKinley, children of Randolph and Mary (Bass) Spalding and grandchildren of Thomas and Sarah Spalding. Charles Bass lived in McIntosh County from the early 1870s until his death in 1919.

250. This operation was located at Kell's Landing on Snow Creek near Lower Bluff, Ashantilly and St. Andrew's Cemetery. See note 224.

251. Catherine Malcolm (ca. 1830-1920), native of Canada. Her husband, John Malcolm (ca. 1840-1890) of Doboy Island, operated a towboat hauling timber from the booms around Darien to the local sawmills. He died from accidental drowning in 1890 when he fell overboard from the tug *John C. Mallonee* near Doboy Island. Sullivan, *Early Days on the Georgia Tidewater*, 468; U.S. Census, 1880, McIntosh County; *Darien Timber Gazette*, July 26, 1890.

252. Charles Carroll Fishburne (1885-1944), native of Barnwell County, South Carolina, and a Darien physician for many years. His wife was Jean (Manson) Fishburne (1899-1947), daughter of John and Bessie (Kenan) Manson of Darien.

253. Townsend at this time was becoming a commercially active community in the western pine woods section of McIntosh County, largely due to the expanding operations of the Georgia Land and Livestock Company (naval stores and cattle) and the Seaboard Air Line Railroad. Sullivan, *Early Days on the Georgia Tidewater*, 580-84, contains an overview of the early 20th century lumber and naval stores industry in west-

ern McIntosh County and how it impacted local communities such as Townsend, Warsaw, Jones, Brickstone and Cox; also, Fuller, Hendrickson and Moon, *Soil Survey of McIntosh County, Georgia,* Series 1929.

254. E. G. Cain (note 232).

255. Legare actually utilized red ink in recording this entry in his Journal, which explains the significance which he attached to this event. In fifty-five years of keeping his Journal in Darien, all the other entries are in black ink.

256. Thomas W. Hardwick, Governor of Georgia, 1921-23.

257. The Dixie Highway was the forerunner of US 17, also called the Atlantic Coastal Highway. It was paved along the Georgia coast in the late 1920s.

258. Pain in the lower back, buttocks, hips or adjacent areas.

259. Charlotte Smith Legare was 74 years of age at the time of her death, being four years older than her husband.

260. Rev. L. E. Williams served the Darien Methodist Church as pastor from 1920-23.

261. Charles Milton Tyson (1862-1940), prominent Darien businessman and financier engaged in the timber and lumber industry and later the management of the Darien Bank. He was president of the Darien Bank from 1915 to 1921 and again from 1922 until his death in 1940. He was married to Emma (Lawson) Tyson (1874-1906). Both are buried in St. Andrew's Cemetery. White, Index to McIntosh County Cemetery Records, 44.

262. Reuben King Hopkins (1872-1939), third son of Octavius C. Hopkins, Sr. (1819-1881) and Elizabeth Aurelia (King) Hopkins (1824-1892). R. K. Hopkins lived at Baisden's Bluff near Crescent. He is buried in the Hopkins Cemetery at Belleville. White, Index to McIntosh County Cemetery Records, 24.

263. Thomas Spalding Wylly (1831-1922), Confederate States army veteran and Darien timberman during the postbellum era. He was an inspector of timber and lumber for Darien in the 1880s and 1890s. He was married to Marion J. Wylly (1832-1917). Both are buried in St. Andrew's Cemetery. White, Index to McIntosh County Cemetery Records, 47; *Darien Timber Gazette,* July 26, 1890; U.S. Census, 1880, McIntosh County.

264. Sophie (LaRoche) Atwood (1851-1930), widow of George E. Atwood (note 221). She is buried in the Atwood family cemetery, Valona.

265. The causeway linking Brunswick with St. Simons Island was completed in July 1924 and marked the beginning of the island's development as a major resort destination. The work on this important project was coordinated by Brunswick engineer Fernando J. Torras. Abby Fuller Graham, *Old Mill Days of St. Simons* (Brunswick, Ga.: St. Simons Public Library, 1976).

266. William J. Hazzard (1858-1928), former McIntosh County merchant and businessman. He managed a store on Doboy Island during the 1880s, later moving his business to Sapelo Sound where the major portion of the timber shipping was concentrated after 1890. Hazzard's ballast island near Creighton Island at the juncture of Front River and Sapelo Sound bears his name, being the site of his store. There was an artesian well on the small island, along with wharfage for the loading of timber and lumber. Sullivan, *Early Days on the Georgia Tidewater,* various references.

267. Margie Hart Legare never married and was fifty years of age at her death.

268. H. Patrick Brannan (1892-1966) was in the naval stores industry and operated a turpentine still north of Darien. He was mayor of Darien, 1928-29, being succeeded by R. Austin Young, Sr., January 1, 1930 (note 188).

269. Ernest Garfield Wilkins (1893-1967) for many years operated a service station on Walton Street in Darien. He was married to Gertrude (Cannon) Wilkins (1900-1966). Both are buried in St. Andrew's Cemetery.

270. Robert Stafford Townsend (1886-1976) worked as a young man at the Hiltons' Lower Bluff sawmill. He and his wife, Ella Elma (Dixon) Townsend (1892-1985) opened the Altamaha Inn on US 17 in Darien in 1929 and operated it for thirty years. Both are buried in St. Andrew's Cemetery.

271. David Edwin Lane (1898-1971) was cashier of the Darien Bank from 1921-1940. He was president of the Darien Bank from 1940-1959 when the bank was located on Broad Street as part of the old Adam Strain building. Sullivan, *The Darien Bank: A Celebration of 100 Years*, 51-56.

272. George Mercer Corlette (1848-1929). White, Index to McIntosh County Cemetery Records, 13.

273. Rufus J. Anderson (1871-1946) was a Darien policeman in the early 1900s, as was his son, Paul Anderson. The Anderson family lived in Darien a block north of the west end of Broad Street on property later owned by the E. B. Kennedy family. Interview with Annie Fisher Gill, March 10, 1997.

274. John Michael Fisher, Jr. (1902-1967), son of John Henry Fisher (1874-1944) and Ann Elizabeth (McCosker) Fisher (1883-1930) of Ashantilly, lived at the Ridge with his wife, Edith (Brewton) Fisher (1900-1960). He was employed by the Georgia Fish and Game Commission. The Fishers later made their permanent residence in Thomasville, Georgia. Interview with Annie Fisher Gill, March 11, 1997.

275. The McIntosh County courthouse has had a history of fires. It was burned as part of the destruction of Darien by federal troops in 1863. Rebuilt after the Civil War, it burned again in early 1873, and was partially damaged in the 1931 fire. The first two fires caused a great loss of county records. Sullivan, *Early Days on the Georgia Tidewater*, 300, 351, 527.

276. This is the final entry recorded by Legare in his Journal before his death four months later on October 14, 1932 at the age of 80.

277. Rev. F. M. Baldwin served the Darien Presbyterian Church as pastor from 1922-1938.

278. Robert Hunter Manson (1893-1970), son of Robert Manson (note 174) of Darien. R. H. Manson was postmaster of Darien in the 1940s and 1950s. He was married to Mary (Sylvester) Manson (1894-1979). Both are buried in St. Andrew's Cemetery.

279. William Southwell Tyson (1897-1972). He was married to Katharine (Parker) Tyson (1900-1954). Both are buried in St. Andrew's Cemetery.

280. Adam Strain Poppell (1875-1950) served a term as receiver of tax returns for McIntosh County in the first decade of the 1900s and was sheriff of McIntosh County from 1920-1948. He was succeeded as sheriff by his son, Thomas H. Poppell, who served in that capacity until his death in 1979. *Darien Gazette*, April 13, 1907.

281. Edd Thompson was a Darien police officer. He also operated a cafe on North Way (US 17) in Darien.

282. James Furse Thomson (1859-1944), native of Barnwell, South Carolina, lived at the Ridge and was married to Caroline (Tyson) Thomson (1876-1957). Both are buried in St. Andrew's Cemetery.

283. H. Wilkes Poppell (1867-1947) was in the naval stores industry and operated turpentine stills in the western pine woods of McIntosh County. He lived at Cox. Interview with A. S. Poppell, Jr., March 6, 1997. White, Index to McIntosh County Cemetery Records, 34.

INDEX

Adams Run, S.C., 3, 8, 25-26, 29
Aiken, Isaac M., 31, 117(n29)
Altama, 123(n64), 131(n122)
Altamaha Woodworking Co., 105, 106, 144(n224)
Anderson, R. J., 111, 149(n273)
Ashantilly, 18, 33, 39, 52, 53, 55, 120(n45), 121(n46), 128(n101), 147(n248)
Asman, B., 137(n166)
Atwood, George E., 97, 110, 140(n190), 144(n221), 148(n264)
Atwood, James A., 80, 141(n196)
Atwood, John M., 92, 143(n213)
Atwood, Sophie L., 110, 148(n264)
Atwood, William H., 31, 46, 74, 93, 118(n31), 139(n180)

Bailey, Arthur, 83, 141(n198)
Bailey, Thomas A., 42, 51, 58, 91, 105, 133(n132)
Baldwin, F. M., 112, 149(n277)
Barclay, Elihu S., 32, 119(n38), 129(n104)
Barclay, Wyatt deR., 42, 109, 129(n104)
Barnwell, Archibald S., 8, 21(n17), 31, 43, 52, 100, 115(n21)
Barnwell, Nathaniel H., 4, 8, 21(n16), 29-30, 88, 113(n1)
Barrington Station, Ga., 48, 49, 131(n121)
Bass, Charles L., 105, 147(n249)
Baxter, James, 73
Bealer, Lewis M., 39, 123(n66)
Bealer, Sarah F., 37, 123(n66)
Belleville, Ga., 41, 126(n91)
Bennett, James R., 116(n27), 138(n170)
Blackbeard Island, 131(n117)
Blackburn, John C., 102, 107
Black Island, 18, 73, 79, 93, 107, 121(n53), 122(n61), 124(n78), 141(n193)
Blount, Evelina M., 30, 38, 114(n10)
Blount, Thomas B., 31, 42, 84, 119(n34)
Blue and Hall, 120(n44), 147(n247)
Bluestein, Hyman, 78, 140(n189)
Bluestein, Jack, 137(n166)
Bluestein, Joseph, 84-85, 140(n189)
Bluestein, Meyer, 65, 95, 100, 111, 112, 137(n166), 140(n189)
Bluestein, Sam, 137(n166)
Bluestein, Walter, 137(n166)
Brannan, H. P., 111, 149(n268)
Brannan, William, 99, 145(n225)
Briesenick, R. E., 72, 73
Britt, Charlotte, see Charlotte B. Massey
Britt, Claudia Legare, 18, 29, 30, 52, 60, 64, 77, 85, 95, 113(n4), 132(n125)
Britt, Emily, see Emily B. Varnedoe
Britt, Jesse A., 50, 52, 60, 77, 82, 85, 100, 107, 132(n125)
Britt, Jesse Eugene, 77, 92, 132(n125)
Britt, John Legare, 69, 70, 82, 85, 92, 109, 111, 132(n125)
Broughton Island, 8, 9, 41, 45, 49, 126(n94)
Brunswick, Ga., 15, 43, 46-48, 72-73, 81, 93, 94, 96, 98, 106, 108, 110, 117(n29), 129(n108), 148(n265)
Butler, G. W., 138(n170)
Butler, Fanny Kemble, see Frances Anne Kemble
Butler, Pierce M., 8, 9, 10, 113(n6), 119(n35), 130(n116)
Butler's Island, 2, 8, 9, 10, 35, 36, 45, 50, 52, 74, 81, 129(n107), 139(n178), 143(n214)

Cain, Edward G., Jr., 84-85, 90, 100, 109, 111, 112, 145(n232)
Cambers Island, 9, 10, 39, 40, 122(n60), 125(n80)
Campbell, Tunis G., 17, 118(n33)
Cathead Creek rice tracts, 8, 39, 74, 118(n32), 120(n41), 122(n54), 124(n75), 131(n120)
Cedar Point, Ga., 74, 132(n124),

139(n180)
Central Railroad Banking, 9, 14,
   21n, 51, 52, 78
Champneys Island, 2, 8, 9, 10, 13,
   14, 21(n17), 43, 51, 52, 53, 55, 56-58,
   60-61, 65-78, 79, 80, 129(n107),
   143(n214)
Churchill, Emma C., 30, 114(n11)
Churchill, W. W., 40, 114(n11),
   141(n197)
Clancy, Raymond W., 100, 112,
   145(n234)
Clark, John, 111
Clark, Peter S., 19, 65, 76, 79, 99,
   100, 104, 133(n133), 136(n162)
Clarke, J. K., 31, 41, 127(n97)
Clifton, William, 90, 142(n211)
Cobb, Sarah, 55, 134(n138)
Coffin, H. E., 23(n43), 143(n216)
Collat, Isidore, 72, 74, 139(n175)
Cook, C. E., 67
Corbin, Richard, 8, 40, 124(n79)
Corlette, George, 111, 149(n272)
Cox, see Barrington Station
Crawford, Lectured, 68, 138(n170)
Creighton Island, 15, 21(n13),
   118(n31), 126(n89), 148(n266)
Crescent, Ga., 41, 77, 83, 86, 126(n90)
Cromley, Catherine, 62, 135(n153)
Cromley, Daniel, 74, 139(n177)
Cunningham, T. M., 9, 51, 65, 71, 78
Curry, Albert B., 31, 34, 63, 116(n22)
Curry, A. D., 33, 42

Darien, Ga., 2-8, 14-19, 30-32, 42-43,
   44, 52, 58-59, 62, 65, 86, 88, 89, 95,
   96, 102, 108, 110, 111, 113(n8),
   124(n69), 127(n97)
Darien & Western R.R., 15, 50, 52,
   69, 71, 79, 126(n89), 132(n127)
Darien Bank, 46, 111, 131(n118),
   148(n261), 149(n271)
Darien Ice Manufacturing Co., 17,
   82, 89, 90-91
Darien Junction, Ga., 50, 82, 84,
   133(n128)
Darien Pilots Commission, 15, 65,

137(n163)
Darien Presbyterian Church, 17-18,
   23(n48), 33-34, 49, 58, 60, 62, 63-64,
   82, 84, 86, 97, 99, 107, 125(n84),
   136(n155), 141(n200)
Darien Shipbuilding Co., 102, 103,
   146(n238)
Darien Short Line R.R., 15, 41,
   126(n89), 132(n127)
Darien Telephone Co., 89, 104,
   146(n243)
Darien Timber Gazette (later
   Darien Gazette), 4, 5, 7, 38, 42,
   103, 119(n35), 124(n70), 128(n102)
Davis, Irvin, 77
Davis, Lucien B., 31, 115(n18)
Dean, James, 74, 139(n179)
Dean, Matthew J., 98, 144(n223)
Dean, Samuel R., 89, 141(n199),
   142(n208)
Delegal, Henry, 16-17, 58
DeLorme, L. E. B., 32, 119(n37),
   147(n247)
Dent, James T., 8, 37, 70, 123(n65)
Dimmock, Albert E., 18, 33, 51,
   58, 65, 133(n133)
Doboy Island, 7, 32, 87, 117(n28),
   119(n40), 121(n52), 147(n251),
   148(n266)
Dolbow, C. A., 80, 141(n195)
Donnelly, Caroline, 80
Donnelly, William J., 42
Downey, R. J., 102, 146(n238)
Durant, Frank, E., 88, 142(n205)

earthquake of 1886, 36-37
Egg Island, 41, 127(n95)
Eulonia, Ga., 41, 62, 69, 96, 125(n88)
Evelyn, 29-30, 113(n1)

Fairhope, Ga., 23(n43)
Faries, George W., 58, 134(n143)
Fishburne, Charles C., 106, 110,
   147(n252)
Fisher, John M., 31, 42, 94,
   128(n101), 149(n274)
Fisher, John M., Jr., 111, 149(n274)
Forbes, J. G., 99, 100, 109, 144(n224)
Fox, R. D., 133(n129)

Fulton, Charles O., 35, 77, 79, 90, 121(n51)
Fynn family, 90

Generals Cut, 21(n16), 49, 140(n183)
Generals Island, 2, 4, 8, 9, 18, 21(n16), 30-31, 32-33, 35, 39, 44, 50, 52, 74, 92, 104, 113(n6), 121(n46), 124(n74), 139(n178)
Georgia Coast & Piedmont R.R., 15, 17, 23n, 83, 86, 92, 93, 94, 95, 96, 101, 104, 105, 106, 108, 142(n203), 143(n215, n217), 144(n219), 146(n241)
Gignilliat, Thomas H., 8, 31, 74, 80, 118(n32),
Gignilliat, W. R., 32, 118(n32), 119(n36)
Girardeau family, 26
Grant Chapel Presbyterian Church, 44, 45, 129(n109)
Grant, John, 30, 38
Grubb, Richard W., 4, 5, 8, 42, 103, 119(n35), 124(n70), 128(n102)
Grundy, Fred, 87, 142(n204)
Grundy, Louise H., 87, 142(n204)
Guyton, Alonzo, 31, 96, 116(n27), 138(n170)

Hackel, Annie B., 137(n166), 143(n216)
Hackel, Emanuel, 95, 137(n166), 143(n216)
Hagan, John, 89, 142(n207)
Hagan, Samuel J., 104, 106, 147(n246)
Hagan, William B., 104, 146(n245)
Hamilton, Arthur S., 4, 18, 19, 26, 30, 62-63, 65, 66, 113(n4)
Hammersmith Landing, 36, 43, 49, 123(n64), 132(n122)
Harris Neck, Ga., 15, 44, 84, 129(n112), 130(n113)
Harris, R. B., 30, 31, 114(n12)
Haynes, Anne Lee, 105, 147(n248)
Haynes, Frances, 105, 147(n248)
Haywood, Sam, 72
Hazzard, W. J., 110, 148(n266)
Hessie , 36, 37, 146(n239)

Hilton-Dodge Lumber Co., 6, 7, 50, 68, 81, 110, 114(n9), 117(n28), 127(n97), 133(n129), 136(n157), 141(n192)
Hilton, Joseph, 6, 31, 117(n28), 133(n129)
Hines, James, 86, 142(n203)
Hird Island, 117(n29), 147(n247)
Hofwyl-Broadfield, 8, 37, 95, 123(n65), 126(n94), 146(n241)
Holmes, Jackson M., 52, 133(n135)
Holmes, James E., 119(n35), 124(n71)
Holmes, James Edward, 31, 38, 41, 124(n71), 127(n97)
Hope, Matilda, 16-17, 58
Hopeton, 8, 123(n64), 124(n79), 131(n122)
Hopkins, Charles H., Jr., 31, 52, 116(n26)
Hopkins, Octavius C., Jr., 32, 40, 46, 58, 86, 125(n82)
Hopkins, Reuben K., 90, 109, 112, 148(n262)
Hopkins, Margaret, 146(n237)
Hopkins, Thomas S., 146(n237)
Hopkins, Tom, 40
Hoyt, Henry F., 30, 31, 114(n14)
Hudson community, 71, 138(n171)
Hunter, Thomas M., 18, 49, 52, 53, 132(n124)
Hunter, William, 58, 127(n97), 134(n143)
hurricane of 1893, 47, 131(n119)
hurricane of 1896, 51-52
hurricane of 1898, 53-55, 134(n137), 134(n140)
Huston, T. L., 21(n17), 143(n214)

Inwood, Ga., 35, 122(n55)

Jackson, Bessie R., 146(n243)
Jackson, C. R., 138(n170)
Jackson, Joseph C., 146(n243)
Jackson, Lewis, 117(n27), 138(n170)
Jackson, Richard V., 146(n243)
Johnstons Station, see Ludowici
Jones Station, Ga. (Jonesville), 33, 73, 77, 87, 120(n43), 121(n51),

153

140(n186)
Joselove (later Josel), I., 65, 137(n165)

Kell's Landing, 78, 105-06, 132(n123), 147(n250)
Kemble, Frances Anne, 8, 119(n35), 129(n107), 130(n116)
Kenan, Spalding, 16, 31, 42, 46, 84, 127(n100)
Key, Benjamin W., 31, 116(n23)
King, Smith, 63, 69, 72, 75
Kirby, John J., 127(n97)
Knights of Pythias, 36, 43, 51, 52
Knox, R. H., 131(n118), 133(n129)
Konetzko, Arthur, 110, 129(n105)
Konetzko, William, 43, 129(n105)

Lachlison, Florence C., 67, 137(n167)
Lachlison, James (1), 115(n19), 117(n30)
Lachlison, James (2), 31, 46, 99, 108, 117(n28,30), 133(n129), 137(n167)
Lachlison, Robert, 31, 115(n19) 117(n30)
Lane, D. E., 111, 149(n271)
Lawton, A. R., 16, 59, 78
Lee, A. P., 100, 145(n231)
Legare family, 25-26, 113(n4)
Legare, Charlotte Smith, 3, 17, 18, 19, 26, 29, 85, 109, 113(n4), 148(n259)
Legare, Claudia G., see Claudia L. Britt
Legare, Dollie (Wells), 102, 104, 106, 146(n240)
Legare, Emma Strain, see Emma L. Whitesides
Legare, Houstoun Johnston, 17, 18, 40, 70, 72, 73, 76, 77, 87, 90, 95, 99, 100, 102, 104, 106, 108, 113(n4), 146(n240)
Legare, Houstoun Johnston, Jr., 104, 146(n240)
Legare, John Edward, 146(n240)
Legare, John Girardeau: family relations, 18, 19, 25-26; early life, 26-28; moves to Darien, Ga., 30,

31; cultivates rice at Generals Island, 8, 31-52; and earthquake of 1886, 36-37; moves to Ashantilly, 39; cultivates rice at Champneys Island, 9, 51-78; and 1896 hurricane, 51-52; and 1898 hurricane, 53-55; views on Darien racial unrest of 1899, 16, 58-59; loss of stepson Arthur Hamilton, 62-63, 65; on Darien Aldermanic Board and McIntosh County Commission, 14-15, 65, 76; and Darien Code of Ordinances, 18-19; and Presbyterian Session, 33-34; and Darien Ice Manufacturing Co., 17, 90-91; views on Darien's decline, 15, 17; final years, 19-20, 109-12; death of, 20, 112, 149(n276)
Legare, Margie Hart, 18, 30, 31, 77, 85, 95, 111, 113(n4), 149(n267)
Leigh, Frances B., 8, 9, 21(n16), 45, 46, 50, 52, 130(n116)
Livingston, C. L., 42, 128(n103)
Long, Harvey T., 84
Long, W. J., 104, 146(n244)
Long Reach, 79, 141(n193)
Lorillard, Pierre, 130(n113)
Lower Bluff sawmill, 7, 15, 21(n15), 53, 79, 81, 117(n28), 141(n192)
Ludowici (Johnstons Station), 60, 79, 106, 135(n149)
Lynn, Lucius R., 56, 60, 63, 68, 134(n141)

McCosker, Charles J., 53, 134(n136), 147(n247)
McFarland, F. H., 68, 82, 100, 102, 138(n168)
McIver, S. W., 138(n170)
McKinley, A. C., 121(n52)
Magwood, Carrie, 53
Magnolia House, 38, 124(n69)
Malcolm, Catherine, 106, 147(n251)
Mallard, Charles O. S., 8, 18, 36, 60, 62, 74, 99, 135(n150)
Mallard, William S., 32, 58, 62, 88, 142(n206)
Mann, F. M., 96, 138(n170), 143(n218)

Mansfield, Eleanor "Pet", 74
Mansfield, Joseph, 31, 41, 47, 51, 52, 74, 77, 93, 125(n86)
Manson, Hugh B., 97, 144(n220)
Manson, John, 109
Manson, Robert, 21(n16), 72, 100, 102, 134(n143), 138(n174), 146(n238), 149(n278)
Manson, Robert H., 112, 149(n278)
Manson, Sadie C., 97, 144(n220)
Massey, Charlotte B., 53, 72, 73, 82, 85, 95, 97, 112, 132(n125)
Mayhall Island, 36, 122(n61)
Maxwell, Philip, 45
Meridian, Ga., 88, 142(n205)
Mew, John P., 89, 95, 142(n210)
Middleton, Porter, 100, 145(n226)
Mifflin, Paul R., 138(n170)
Mifflin, Robert, 50
Milne, William S., 85, 86, 99, 142(n201)
Morris, R. L., 35, 122(n54)
Motte, John W., 42, 127(n99)
Muller, Ann (King), 44, 129(n112)
Muller, John, 44, 84, 129(n112)

Nanias, Fortunato, 65, 137(n164)
Nightingale, William, 36, 40, 122(n60)

Oak Hill, see Cedar Point

Palmer, Julia F., 42
Parnell, F. G., 69
Patelidas, George, 100
Paul, R. P., 51, 133(n129)
Pease, Augusta P., 86, 114(n15), 142(n202)
Pease, T. P., 31, 114(n15), 126(n92)
Pine Harbor, see Fairhope
Pinkerton, Samuel J., 31, 116(n24)
Poppell, Adam S., 112, 149(n280)
Poppell, Edward R., 46, 75, 139(n181)
Poppell, George W., 100, 145(n228)
Poppell, H. W., 112, 150(n283)
Poppell, Lowndes, 55, 134(n138)
Presbyterian Church, see Darien Presbyterian Church

Pumpkin Hammock, see Union Island

Quarterman, C. M., 31, 44, 115(n20)
Quarterman, T. S., 31, 32, 42, 115(n17)

racial unrest of 1899, Darien, 16-17, 58-59
racial unrest of 1900, Darien, 62-63
Ravenel, H. S., 32, 46, 58, 134(n142)
Rees, Henry K., 45, 130(n115)
Rhett's Island, 36, 39, 44, 48, 58, 76, 81, 122(n56)
rice industry, 1-2, 4, 8-14, 35, 39, 41, 47, 50-52, 56-58, 60-61, 65-78
Ridge (Ridgeville), 33, 58, 78, 83, 90, 104, 120(n42), 147(n247)
Rogers, Claudia W., 87, 112, 141(n199)
Rogers, W. H., 138(n170)
Rothschild, Charles, 31, 35, 121(n50)
Roughan, Mary B., 104, 146(n242)

St. Andrews Episcopal Church, 31, 45, 103, 130(n115)
St. Cyprians Episcopal Church, 52, 96, 138(n170), 143(n218)
Sapelo Island, 23(n43), 74, 105, 106, 121(n52), 136(n153), 139(n177), 147(n249)
sawmills, see timber industry
Schmidt, Carl A., 31, 91, 96, 127(n97), 131(n118), 143(n212)
Shaw, Robert G., 123(n67)
Shell Bluff, see Valona
Shell Bluff Canning Co., 78, 79, 140(n190)
Shine, Ann H., 78, 140(n187)
Sidon, 48, 131(n120)
Sinclair, B. T., 120(n41)
Sinclair, David S., 8, 32, 33, 48, 56, 73, 74, 89, 120(n41), 131(n120)
Sinclair, Lillian C., 142(n210)
Sinclair, Lillian F., 14, 73, 120(n41)
Sinclair, W. W., 73, 80, 120(n41), 142(n210)

Smith, N. Keff, 18, 40, 41, 45, 68, 82, 84, 99, 109, 125(n84)
Snows Creek, 39, 124(n78)
Space, Julian A., 100, 145(n230)
Spalding, Thomas, 120(n45), 121(n52), 126(n92)
Spalding, Thomas B., 35, 121(n52), 128(n103)
Stebbins, James L., 100, 145(n236)
Stebbins, John S., 41, 127(n96)
Stokely, Aleck, 55, 134(n138)
Strain, Adam, 30, 31, 33, 53, 113(n8), 131(n118)
Strain, Richard H., 21(n17), 76
Strain, Robert A., 8, 21(n16), 79, 84, 91, 98, 110, 140(n190), 141(n194)
Strain, William H., 8, 21(n17), 74, 76, 78, 84, 131(n118), 139(n176)

Taylor, J. D., 44, 138(n170)
Terrell, A. C., 107
Thicket, 41, 47, 114(n15), 126(n92)
Thomson, James F., 112, 150(n282)
Thompson, Edd, 112, 150(n281)
Thorpe, E. M., 131(n118)
Threemile Cut, 21(n16), 75, 140(n183)
timber industry, 3-8, 15-16, 21(n8, 13, 15), 125(n81), 127(n97), 134(n140), 134(n143), 136(n157), 136(n158)
Todd, Henry, 36, 38, 44, 122(n57)
Tompkins, Henry B., 31, 118(n33)
Townsend, Ga., 104, 107, 138(n172), 147(n253)
Townsend, James S., 49, 132(n123)
Townsend, Joseph E., 16, 58, 135(n147), 138(n172)
Townsend, Robert S., 111, 149(n270)
Townsend, William R., Sr. and Jr., 59, 132(n123), 135(n147)
Tyson, Charles M., 109, 112, 131(n118), 146(n238), 148(n261)
Tyson, William S., 112, 149(n279)

Union Island, 7, 55, 105, 147(n247)
Upper Mill, 36, 80, 122(n58)
Valona, Ga., 78, 92, 93, 97, 110, 140(n190), 144(n221)

Varnedoe, Emily B., 60, 85, 95, 97, 132(n125), 144(n222)
Varner, Paul, 111

Walker, Charles R., 18, 58, 70, 83, 135(n144)
Walker, David S., 100, 145(n235)
Walker, James, Jr., 21n, 31, 36, 42, 48, 56, 115(n16), 116(n25), 127(n98)
Walker, Joseph A., 42, 43, 44, 90, 115(n16), 116(n25), 127(n98)
Walker Reuben K., 31, 41, 80, 99, 115(n16), 126(n89)
Warsaw, see Darien Junction
Way, S. A., 42, 63-64, 136(n161)
Weil, Henry A., 41, 42, 51, 87, 125(n87)
Weil, Simon A., 46, 125(n87)
Whitesides, Claudia, see Claudia W. Rogers
Whitesides, Daniel L., 84, 85, 87, 141(n199)
Whitesides, Emma Legare, 18, 19, 35, 70, 72, 84, 85, 87, 105, 107, 110, 112, 113(n4), 141(n199)
Whitesides, Joseph L., 85, 86, 112, 141(n199)
Wilcox, William A., 81, 114(n11), 141(n197)
Wilkins, Ernest G., 111, 112, 149(n269)
Williams, L. E., 109, 148(n260)
Williams, Sidney W., 89, 142(n209)
Wister, Sarah B., 9, 21(n16), 130(n116), 144(n218)
Wolf Island, 44, 55, 130(n114), 134(n138)
Wylly, Alexander C., 32, 46, 64, 89, 136(n161)
Wylly, Thomas S., 109, 148(n263)
Wylly, William C., 8, 14, 21(n17), 41, 78, 106, 109, 126(n93)

yellow fever, 46-48, 131(n117)
Young, Robert A., Sr., 67, 78, 100, 111, 140(n188), 149(n268)
Young, Robert A., Jr., 111, 140(n188)
Young, W. M., 42, 128(n101)

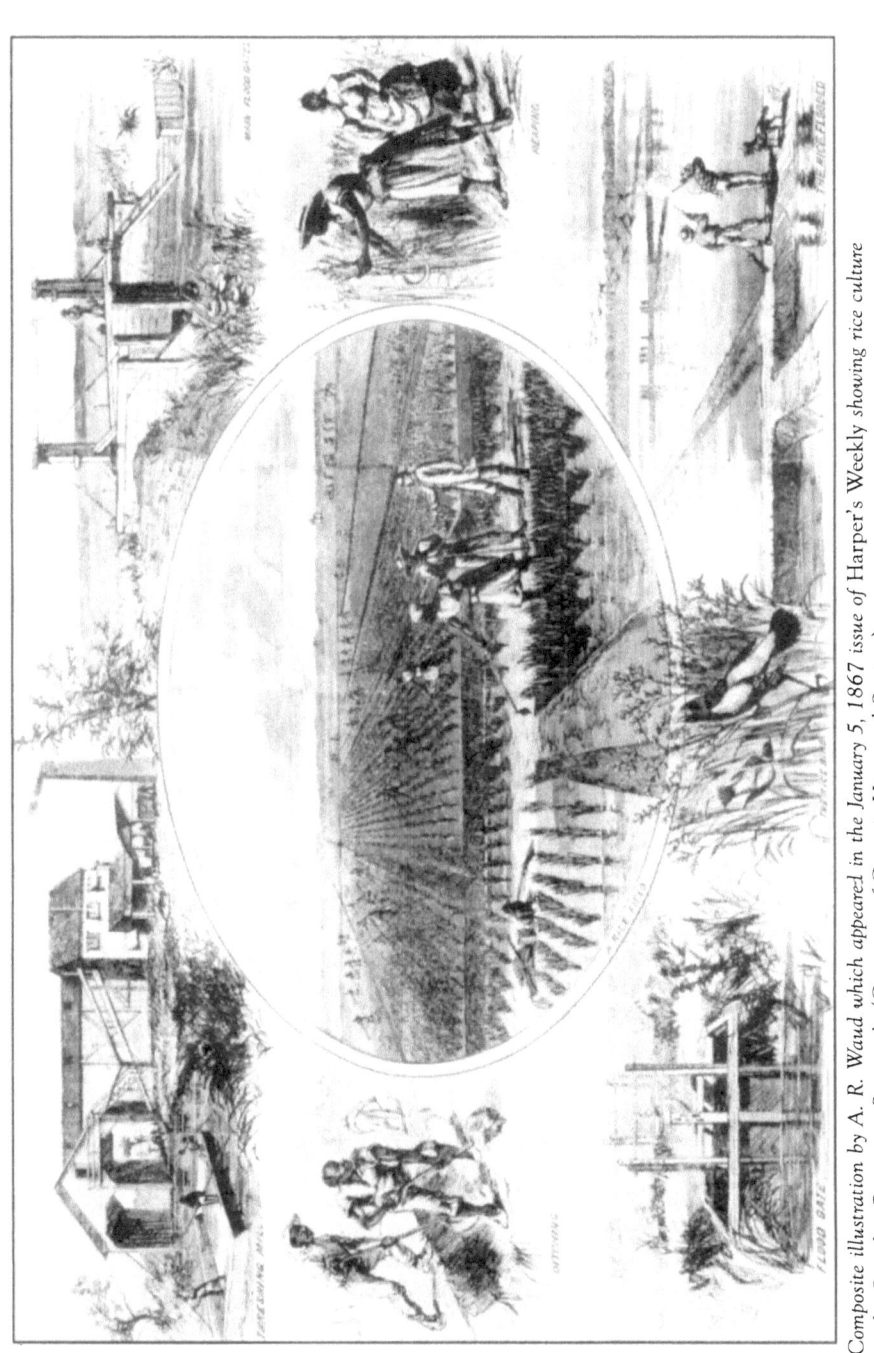

Composite illustration by A. R. Waud which appeared in the January 5, 1867 issue of Harper's Weekly showing rice culture on the Ogeechee River near Savannah. (Courtesy of Georgia Historical Society).

www.ingramcontent.com/pod-product-compliance
Ingram Content Group UK Ltd.
Pitfield, Milton Keynes, MK11 3LW, UK
UKHW041451180426
11946UKWH00013B/154/J